ORCHIDS OF VANUATU

T0136789

ORCHIDS of VANUATU

by B. Lewis & P. Cribb

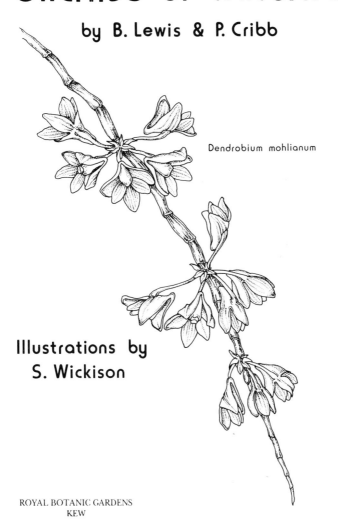

Dendrobium mohlianum

Illustrations by S. Wickison

ROYAL BOTANIC GARDENS
KEW

General editor of Series M.J.E. Coode. Special editor for this volume Joyce Stewart. Cover design by Sue Wickison, front cover depicting a wooden drum and back cover depicting a totem figure made of tree fern, both from Ambrym. The orchid is *Spathoglottis petri*. Text set by Pam Arnold, Christine Beard, Margaret Newman, Helen O'Brien and Pam Rosen.

ISBN 0 947643 16 8

Printed in Great Britain by
Whitstable Litho Printers Ltd

For G. Hermon Slade

FES TOK

Ating olgeta man we oli ridim ripot ia bambae oli intres blong save olsem wanem hem i stat. Taem mitufala i staf raetem ripot ia, i gat ol ripot tu we oli stap priperem long Fiji, Ostrelia mo long ol aelan blong Malaya abaot ol kaen plan we oli kolem OKID. Long olgeta eria klosap, Niu Kaledonia nomo i bin gat wan gudfala ripot we oli raetem i no longtaem abaot ol flaoa blong hem (Hallé, 1977). Vanuatu mo Solomon Aelan long not, tufala i stap stret long medel blong rijen ia we oli stap mekem plante stadi long hem be oli no save plante samting abaot ol okid we oli save faenem long ol grup blong aelan klosap.

Afta long ol strong tingting mo sapot blong mane we i kamaot long G. Hermon Slade mo Okid Faondesen blong Ostrelia, wok i bin gohed long Kew long 1980 blong traem blong mekem situesen i kam gud. Ol ripot blong tu impoten seksen blong *Dendrobium* long saed blong ol plan, we i gat plante long Vanuatu, oli bin raetem finis (Cribb, 1983, 1986) wetem wan fesfala list blong ol okid long Solomon Aelan mo Bougainville (Thorne and Cribb, 1984). Ripot ia 'Orchids of Vanuatu' hem i nekis step blong wan risej program we i no longtaem bambae i givim frut blong hem wetem wan buk we singaotem 'Orchids of the Solomon Islands and Bougainville'.

Mifala i sua se i gat plante plan yet we man i neva fainem long Vanuatu mo ol studi long fiuja bambae i soem ol niu rikod blong ol okid plan, ol ples we i gat ol defren kaen okid long hem mo tu bambae i save soem ol niu okid plan. Mifala i hop se buk ia bambae i soem stret wei we Vanuatu hem i rij long okid mo bambae i soem tu ol eria we i gat ol defren kaen plan long hem we i gud blong kaontri i protektem. Olgeta okid blong Vanuatu i wan impoten samting blong kaontri mo, sipos yumi protektem, bambae i save mekem se plante turis i kam long kaontri blong luk ol wael flaoa blong ol aelan wetem ae blong olgeta.

Mifala i glad tumas long strong tingting mo help we Gavman blong Vanuatu mo Ministri blong Naturol Risos i bin givin long mifala. Mifala i hop se bambae yufala i akseptem buk ia olsem wan saen blong talem tankiu from wom welkam we mifala i kasem long olgeta pipol blong Vanuatu.

PREFACE

The genesis of this account may be of interest to readers. At the time of writing, floristic accounts of the orchids are in preparation for Fiji, Australia and the Malay Archipelago. Of adjacent areas, only New Caledonia is served by a recent critical floristic account (Hallé, 1977). Vanuatu, and the Solomon Islands to the north, lie in the centre of this Pacific region but little is known of their orchid floras or of the distribution of their orchid species within the adjacent groups of islands. Encouraged and financially supported by G. Hermon Slade and the Australian Orchid Foundation, work began at Kew in 1980. Accounts of two of the horticulturally important sections of *Dendrobium*, both well represented in Vanuatu, have been published already (Cribb, 1983, 1986) together with a preliminary checklist of the orchids of the Solomon Islands and Bougainville (Thorne & Cribb, 1984).

This account, 'Orchids of Vanuatu', represents the next stage of a research programme that will culminate soon with the publication of 'Orchids of the Solomon Islands and Bougainville'.

This is the first comprehensive account of the orchids of Vanuatu and includes 16 new genera and 53 new species records. We are sure that much still remains to be discovered in Vanuatu and that future exploration will reveal new records of orchids, extensions of range of known species, and even new species. We hope that this book will demonstrate the richness of Vanuatu's orchid flora and will pinpoint areas of species' diversity that might be considered for special protection. The orchids of Vanuatu are a national asset and, if protected, could attract visitors to the islands.

We are grateful for the encouragement and help given to us by the Government of Vanuatu and the Ministry of Natural Resources. We hope that this publication will be accepted as a small token of gratitude for the warm welcome we have received from the people of Vanuatu.

CONTENTS

LIST OF MAPS AND FIGURES

All by Sue Wickison (except for fig. 11 by Sarah Robbins).

INTRODUCTION

Vanuatu, formerly the New Hebrides, consists of 80 islands, strung out in the shape of a figure Y 800 km long. They lie in the south-west Pacific Ocean between latitudes 12-21°S and longitudes 166-170°E (map 1). Until recently they were called the New Hebrides because Captain James Cook, who explored the islands in 1774, thought that they looked like the Scottish Hebrides. The New Hebrides were ruled jointly by Britain and France as a condominium until their independence in 1980, when the islands were renamed Vanuatu, which means 'our land'.

Of these 80 islands, only about 12 are significant in human and economic terms. The area of each of the principal islands is as follows:

Espiritu Santo	3885 sq.km	Pentecost	324 sq.km
Malekula	1166 sq.km	Epi	259 sq.km
Erromango	1165 sq.km	Maewo	233 sq.km
Efate	777 sq.km	Anatom	159 sq.km
Ambae	420 sq.km	Vanua Lava	80 sq.km
Ambrym	414 sq.km	Gaua	80 sq.km
Tanna	390 sq.km	Torres group	78 sq.km

The names of the islands in current usage and obsolete alternatives are given in appendix 1.

The islands cover an area of 14,763 sq.km. The highest point is Mt. Tabwemasana (1889 m) on Espiritu Santo. Most of the islands are volcanic, some actively so.

The northern region of the Vanuatu archipelago contains the Torres Islands, the Banks Islands (Vanua Lava and Gaua), Espiritu Santo (usually called Santo), Malekula, Maewo, Ambae, Pentecost, Ambrym, Epi and the Shepherd Islands.

The southern region of the archipelago consists of four islands of very different size, clearly separated from and with a cooler climate than the remainder of the archipelago (Schmid, 1975). The capital, Port Vila, is on Efate, the northernmost island of the group. Anatom and Erromango are sparsely populated and still possess extensive climax vegetation on their deep, infertile soils. Tanna has a dense human population and scarcely any primary vegetation remains there. The soil is generally fertile and is periodically rejuvenated by ash showers from a continuously active volcano, Yasur (Schmid, 1975).

Physical Geography

The main islands can be considered in three geological groups (Mawson, 1905): islands in which extensive outcrops of Miocene volcanic and sedimentary rocks occur (Espiritu Santo, Pentecost and Malekula); the remaining volcanic islands, composed of more recently extruded materials (Banks Islands, Ambrym, Efate, Epi, Lopevi and Tanna); and the small islands formed almost entirely of coral limestone (map 2).

The islands were formed by a fold-ridge, apparently continuous with that passing round the north of New Guinea, through Sumatra, to the Himalayas and

1

Map 1: Position of Vanuatu in the south-west Pacific.
after N. Douglas, 1986.

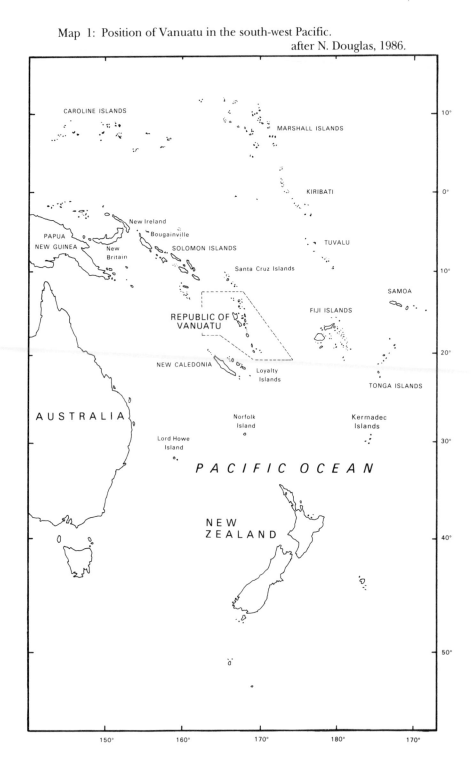

Map 2: Physical geography of Vanuatu.

after Mallick, 1975.

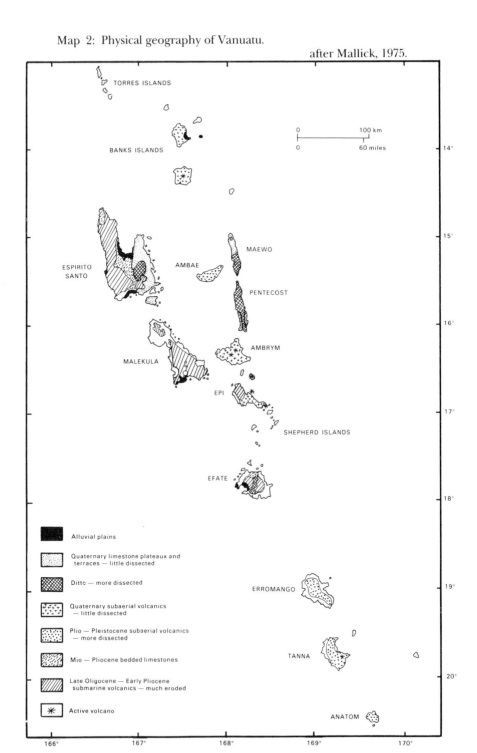

southern Europe. This upfold would appear to have defined the line of the present western islands, in which extensive outcrops of Miocene strata occur. With faulting there was an extrusion of basalt and andesite lavas and subsidence of the sea floor to the east.

The volcanic rocks of Vanuatu are of varied composition, including basalts, andesites and dacites. In some areas, such as Epi and Efate, two or all three of these types occur in close proximity to one another. In others, one type is largely predominant. Thus Pentecost and Maewo in the north, appear to be formed mainly of basalts, and Tanna and Erromango, in the south, of andesites. This difference, indicating primarily variation in silica content, is of considerable importance owing to its influence on the composition of the soils of the islands.

There are active volcanos on the islands of Ambrym, Tanna and Lopevi (between Ambrym and Epi). The island of Ambrym represents the truncated cone of an ancient volcano of great size. The floor of the old crater is now thickly covered with ash from the outbursts of daughter cones formed within it. The most recent violent eruptions occurred in October 1894, December 1913, June–July 1929 and March–April 1937. The 1913 eruption was accompanied by great changes in the level of the land. In some places the coastal region was greatly raised, so that the coastline shifted appreciably. Elsewhere there was considerable subsidence.

Volcanic activity on the islands of Lopevi and Tanna has been much less violent. Lopevi is a small island consisting almost entirely of a volcanic cone. It was reported to have been very active in 1863–4. Since that time there have been several further eruptions, the most recent in November 1939.

The volcano on Tanna, Yasur, is continuously active, but it causes the inhabitants of the island no inconvenience. There are other indications of volcanic activity at many places in the archipelago. Hot springs can be found on the Banks Islands, Pentecost, Ambrym, Epi, Efate and Tanna, and escapes of gas on Tongoa in the Shepherd Islands.

It is thought that the Shepherd Islands, which are a group of seven small islands together with numerous islets, reefs and banks, were once a large island named Kuwae which was subjected to volcanic eruptions and earthquakes in the late fifteenth century. The Shepherd Islands are all of recent volcanic origin.

In many areas the volcanic rocks are overlain by coral limestone, where successive uplifts of the land have raised former fringing reefs above sea-level. Among the most important coral areas are those in Espiritu Santo and Malekula.

The many small, flat islets off the coasts of the major islands have been similarly formed by a combination of volcanic and coralliferous material. Some, such as Rano and Atchin, off the coast of Malekula, are primarily volcanic with only a thin coral crust, while others, for example Lelepa and Eradaka, off the west coast of Efate, appear to be built up very largely of coral limestone.

Climate

The climate varies from tropical in the north to sub-tropical in the south. Temperature and rainfall in the Banks and Torres Islands and the northern half of Vanuatu resemble those of the Solomon Islands. In the southern islands conditions are less tropical with considerably greater seasonal variation in temperature, and lighter rainfall. Average maximum temperatures are 30°C from January to March and 25°C from July to August. Average minimums for those months are 23°C and 19°C.

Rainfall is heavier on the northern islands, with the Banks islands getting 4000 mm of rain a year, and Espiritu Santo 3100 mm, compared with the group's mean annual rainfall of 2300 mm. Rainfall is also heavier on the higher islands and upon the windward sides of the islands.

Vanuatu lies in the region of the south-east trade winds which predominate throughout the year. However, between the beginning of November and the middle of April, north-easterly winds tend to bring rain and during the first quarter of the year the northern part of the group may experience cyclones. On several occasions in recent years they have caused great damage, the most devastating being Cyclone Uma, which struck Port Vila in February 1987.

Vegetation of the Islands

The predominant vegetation of the islands is tropical rainforest, floristically related to that of New Guinea and the Solomon Islands, but with fewer families, genera and species represented (Chew, 1975). In fact the Torres Islands, Vanuatu's northernmost group, are only about 100 km away from the Santa Cruz Islands, the southernmost in the Solomon Islands.

Extensive coastal forest with *Casuarina, Hibiscus* and *Pandanus* can be found in some places. On Efate, there are swamp forests with *Barringtonia* and *Pandanus* and also scattered mangrove forests. The tropical lowland evergreen rainforest behind is dominated by *Castanospermum, Euodia* and *Hernandia*. Small areas of broad-leaved deciduous forest and closed conifer forest, dominated by *Agathis* spp., are restricted to western parts of Espiritu Santo, Erromanga and Anatom. Montane rain forest occurs between 1000–1500 m and cloud forest, with *Metrosideros*, above 1500 m.

Cyclones have had a profound effect on the vegetation of the islands. Given the past and present impact of man and cyclones on the forest, it is reasonable to assume that climax forest can only exist on protected, inaccessible sites. The bulk of the forest therefore is of secondary status. The Vanuatu archipelago extends through a range of latitudes, and thus varying seasonality contributes to a recognizable difference between the floras of the northern and southern islands (Chew, 1975). This difference is expressed in many ways, some species are important and frequent in the northern islands but occur only infrequently in the southern islands (Chew, 1975). Other plants that occur mainly at higher altitudes in the northern islands also occur abundantly at lower altitudes in the southern islands. Chew further suggests that the northern islands are much more diversified and have more Indo-Malesian elements than the southern islands. However, the higher diversity in the north is not entirely due to climate; area and altitude are important co-determinates since Espiritu Santo and Malekula are the two largest islands in the archipelago. The Vanuatu archipelago thus differs from the Solomon Islands in that there are significant differences in the flora of the different islands. These differences arise because of the diversity of size, height and climate of the islands.

Early in the nineteenth century British traders visited the islands for sandalwood (*Santalum* spp.), much prized in China because of its sweet smell. This trade has continued up to the present time, resulting in a serious depletion of this resource (Bellamy and Saunders, 1987). Extensive areas of forest have also been cleared for plantations and pastures, particularly on the plateaux of Efate and Espiritu Santo, where yams, manioc and bananas are grown for local use. Around the coast, plantations have been established to grow copra, cocoa and coffee and also to raise cattle, while forestry and fishing are also important in some areas. However, Erromango still has about 180 sq.km of closed forest (Schmid, 1978) and at present the Vanuatu government and the landowners are proposing to establish a kauri (*Agathis* spp.) reserve on Erromango (Gillison and Neil, 1987). A botanical garden at Tagabe, on Efate, is also proposed.

Surveys of the Vanuatu flora have been undertaken by A. Guillaumin (1948) and M. Schmid (1973); while S. Gowers (1976) has catalogued the common trees of Vanuatu. The government is currently embarking on a national forest resources survey/inventory, but, at present, no comprehensive floristic account of Vanuatu exists.

The Origins of the Vanuatu Flora

The Vanuatu flora has been influenced by migrations from the Indo-Malaysian region and to a lesser extent from New Caledonia.

Chew (1975) suggested that Vanuatu has a young immigrant flora and that the colonization of the archipelego has taken place very rapidly in recent times. Therefore the flora has had insufficient isolation in space and time to permit speciation to develop to any significant degree.

Plant dispersal to the Vanuatu archipelago is likely to have been transmarine rather than by migration over land. Vanuatu is oceanic in origin (Mallick, 1975) and probably had no land connection with any of the neighbouring archipelagos since its origin. Vanuatu is not only very young but Mallick concludes that about 90% of its present area became aerial (and thus available for colonization by land plants) only in the last 1.5 million years, hence the colonization of the islands has been extremely rapid. Whitmore (1973) explained Pacific plant geography in terms of plate tectonics. He argued that in the Fiji region there has been an overall movement of the islands eastwards away from the plate on which Australia lies. Tonga and Vanuatu are believed to have been a single island arc in the Oligocene, now Tonga is 15°E of Vanuatu because of complex rotation and the spreading and eastward movement of the ocean floor (Karig, 1970). The overall effect is that the islands of the south-west Pacific are much further apart from each other and from Australia now than in the early Tertiary. However, it does not seem necessary to invoke plate tectonics to account for Pacific plant geography. Crustal plates have indeed moved and spread, but these events occurred much too early to have any overall effects on the distribution pattern of the present-day flora of this region.

Botanical Exploration of Vanuatu
(For a summary see appendix II).

George Forster, the botanist accompanying Captain Cook in the Resolution, collected the first plants from Vanuatu in 1774 from the islands of Tanna and Malekula. These included the saprophytic orchid, *Dipodium punctatum* var. *squamatum*.

In 1853 HMS Herald visited Vanuatu and the naturalists John MacGillivray and William Milne collected orchids, including the spectacular *Dendrobium macrophyllum* and *Megastylis gigas*, from Anatom. John MacGillivray returned to Vanuatu in 1859 when he collected the type specimen of *Peristylus stenodontus*.

Before 1881 (date not traced) Captain George Braithwaite collected orchids from Efate and Erromango which were later studied by Schlechter and Kraenzlin. His collections include the type of *Dendrobium psyche* (a synonym of *D. macrophyllum*).

Between 1872 and 1873 the Australian missionaries Rev. Frederick A. Campbell and Rev. Fraser collected plants for Baron Ferdinand von Mueller, the Director of the Melbourne Botanic Gardens and in 1873, nearly a hundred years after Cook visited and named the New Hebrides, Mueller published the earliest account of its flora. In this he listed only four orchids: *Spathoglottis pacifica*, *Corymborkis veratrifolia*, *Peristylus novoebudarum* and *Gastrodia orobanchoides* (a synonym of *G. cunninghamii*).

In 1883, David Levat collected orchids from Efate. Many of these collections were designated as types by Kraenzlin, who described and identified many of the orchids for Guillaumin's work on the flora of the New Hebrides. These include the type of *Pholidota grandis* (a synonym of *P. imbricata*).

In 1889, Admiral Fairfax, Commander of H.M. Fleet in Torres Straits, collected plants, including *Dendrobium fairfaxii* which is synonymous with the endemic *D. mooreanum*.

At the suggestion of Mueller, Dr. Alexander Morrison, botanist of the Western Australian government, visited Vanuatu from June to August 1896. His collections included many orchids from Anatom, Erromango and Efate, notably the type species of *Corybas mirabilis*.

W.T. Quaife collected orchids from Espiritu Santo in 1903 including the types of *Dendrobium quaifei* (a synonym of *D. mooreanum*) and *Sarcanthopsis quaifei* (a synonym of *S. nagarensis*).

Many of the missionaries in the early nineteenth century collected plants sending them back to Australia. These collectors include Rev. Daniel MacDonald (1872–1906), Rev. Jos. H. Lawrie (c. 1905) and Rev. A. William Gunn (1915).

The first High Commissioner for the Western Pacific, Sir Everard Im Thurn, collected orchids from around Port Vila, Efate, in 1906, including the type of *Dendrobium sladei*.

Between 1928 and 1929, S. Frank Kajewski collected plants in the south-west Pacific for the Arnold Arboretum. He visited the islands of Anatom, Tanna, Erromango, Efate and Vanua Lava in the Banks group. These collections were studied in 1932 and 1933 by Oakes Ames who described many of the species in his honour, for example *Dendrobium kajewskii* (a synonym of *D. conathum*).

From 1928–1930 Miss Lucy Evelyn Cheesman travelled throughout the Pacific collecting insects and plants for the British Museum. She spent two years in Vanuatu and returned in 1954–1955. She collected orchids from the Banks Islands, Malekula, Erromango and Anatom, including *Vrydagzynea cheesemanii* (a synonym of *V. salomonense*) and *Dendrobium critae-rubrae* (a synonym of *D. austrocaledonicum*).

From October 1933 to February 1934 Oxford University mounted an expedition to Vanuatu, led by John R. Baker, to climb the highest mountain in Vanuatu, Mt. Tabwemasana. On this expedition a number of orchids were collected by his sister Ina Baker assisted by Zita Baker.

In February 1934 the French geologist Edgar Aubert de la Rüe visited Pentecost, Ambrym, Espiritu Santo, Erromango, Epi and Ambae. He returned in October 1935 to June 1936 with Mme. E. Aubert de la Rüe and collected from Epi, Ambrym, Pentecost and Erromango. Many of these collections were first studied by the French botanist Guillaumin who worked on the flora of Vanuatu between 1919–1956.

Luciano Bernardi, a Swiss botanist, collected on Vanuatu from April to June 1968, from the islands Tanna, Erromango and Anatom.

More recently, noteworthy collections have been made by The Royal Society and Percy Sladen Expedition to the New Hebrides in 1971. This expedition visited Espiritu Santo, Malekula, Efate, Erromango, Tanna, Anatom and some small offshore islands adjacent to these (see map 3). Orchid collections were made by Peter S. Green, Nicolas Hallé, Chew Wee-Lek, Jean Raynal, Maurice Schmid and Jean-Marie Veillon.

In 1974–5 H. Bregulla collected orchids from Efate, Espiritu Santo and Tongoa (Shepherd Islands).

From 1975 onwards G. Hermon Slade, a resident of Port Vila, Efate, has been active by collecting and cultivating native Vanuatu orchids. Jacques Begaud and Gabriel Cayrol, in Noumea, have also cultivated Vanuatu orchids. The former collected and flowered the type of *Cadetia quadrangularis*.

From the late 1970s onwards the French organization ORSTOM has been active in Vanuatu, and many important collections have been made by Pierre Cabalion, Geneviève Bourdy, Jean-Marie Veillon, Bernard Suprin, Maurice Schmid, Siri Seoule and Chanel Sam. In 1974 J. Hooke, working for ORSTOM in Noumea, produced a provisional orchid flora for Vanuatu, which listed 83 species in 34 genera, however his manuscript has not been published and many of the orchids were named only to genus.

Between 1976 and 1983 Philippe Morat collected from Pentecost, Espiritu Santo, Efate, Tanna, Vanua Lava and Aniwa.

In 1982 and 1983 Siri Seoule collected from Efate and Anatom.

The orchid account in Flore de la Nouvelle-Calédonie et Dépendances by Nicolas Hallé (1977) included many orchid species occuring in Vanuatu.

In 1980 Phillip Cribb and Alasdair A.O. Morrison collected orchids from Efate. Phillip Cribb returned in 1988 and with Jos. Wheatley collected 92 species of orchid on Espiritu Santo, adding many new generic and specific records for the islands. Jos. Wheatley, a forest botanist in Vanuatu, has continued to collect orchids, and has added many new records especially from the relatively unexplored northern islands of Vanua Lava, Ambae and Pentecost.

The Orchids

The Orchidaceae form the major component of the epiphytic flora of Vanuatu. Chew (1975) estimated that there were 31 endemic species of orchid (based on the Royal Society Expedition results, 1971), but this estimate is undoubtedly rather high. Vanuatu has no endemic genera, although Ames (1932) described *Trichochilus* as a new genus based on *Trichochilus neo-ebudicus* from Erromango. However, this taxon has since been reduced to synonymy in *Dipodium*. Ames also described many new orchid species from Vanuatu in 1932, but admitted his intentional emphasis on differences rather than on similarities

Map 3: Areas visited by the 1971 Royal Society Expedition to the New Hebrides.
after Lee, 1974.

numbers in ( ) represent the number of species reported from the island to date (1989)

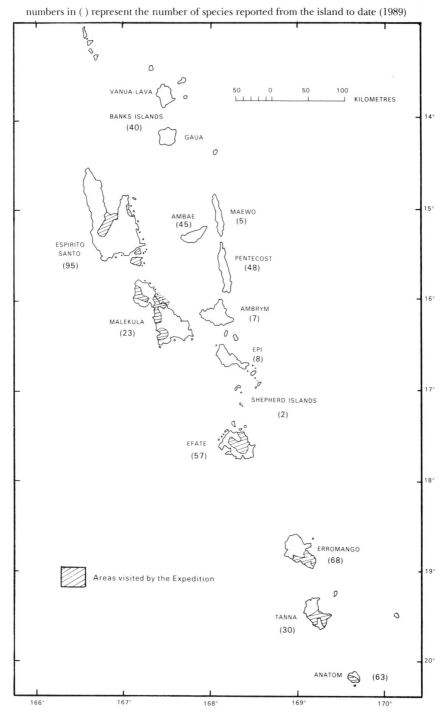

VANUA-LAVA

BANKS ISLANDS
(40)

GAUA

50   0   50   100
|⎯⎯⎯⎯⎯⎯⎯⎯⎯⎯⎯⎯⎯⎯⎯⎯⎯⎯⎯| KILOMETRES

14°

AMBAE
(45)

MAEWO
(5)

ESPIRITO
SANTO
(95)

PENTECOST
(48)

15°

MALEKULA
(23)

AMBRYM
(7)

16°

EPI
(8)

SHEPHERD ISLANDS
(2)

17°

EFATE
(57)

18°

ERROMANGO
(68)

Areas visited by the Expedition

19°

TANNA
(30)

20°

ANATOM   (63)

166°   167°   168°   169°   170°

Table 1: Survey of the Epiphytic and Terrestrial Genera and Species in Vanuatu

Epiphytic		Terrestrial	
Genera	**No of species**	**Genera**	**No of species**
Aglossorhyncha	1	Acanthephippium	1
Agrostophyllum	5	Calanthe	3
Appendicula	3	Cheirostylis	1
Bulbophyllum	17	Corybas	3
Cadetia	1	Corymborkis	1
Ceratostylis	1	Chrysoglossum	1
Cleisostoma	1	Cryptostylis	1
Coelogyne	2	Cystorchis	1
Dendrobium	28	Didymoplexis	1
Diplocaulobium	1	Dipodium	1
Earina	1	Epipogium	1
Epiblastus	1	Erythrodes	2
Eria	1	Eurycentrum	1
Flickingeria	1	Eulophia	1
Glomera	2	Gastrodia	1
Glossorhyncha	1	Geodorum	1
Gunnarella	1	Goodyera	3
Liparis	5	Habenaria	1
Luisia	1	Hetaeria	1
Mediocalcar	2	(Liparis	1)*
Microtatorchis	1	Macodes	1
Oberonia	4	Malaxis	4
Octarrhena	1	Megastylis	1
Pedilochilus	1	Microtis	1
Pholidota	1	Moerenhoutia	1
Phreatia	6	Nervilia	2
Pomatocalpa	1	Neuwiedia	1
Robiquetia	1	Oeceoclades	1
Sarcanthopsis	1	Peristylus	5
Schoenorchis	1	Phaius	3
Taeniophyllum	1	Pristiglottis	1
Thrixspermum	2	Spathoglottis	4
Trichotosia	1	Spiranthes	1
Tuberolabium	1	Tropidia	1
		Vrydagzynea	3
		Zeuxine	2

GENERA = 34   **SPECIES** = 99   **GENERA** = 35   **SPECIES** = 59

TOTAL NO. OF GENERA = 69

TOTAL NO. OF SPECIES = 158

* genus contains epiphytic and terrestrial species

as he felt that such an action would be less detrimental to biogeographical science! As the wide distribution of many Pacific Island orchids has become understood, many names have been reduced to synonymy.

In this checklist seven endemic species are listed: *Peristylus stenodontus*, *P. wheatleyi*, *Liparis aaronii*, *Cadetia quadrangularis*, *Dendrobium greenianum*, *D. mooreanum* and *Pedilochilus hermonii*.

Approximately two-thirds (62%) of the species are epiphytic with a third (38%) terrestrial (see table 1), four of these being saprophytic species, *Gastrodia cunninghamii*, *Dipodium punctatum* var. *squamatum*, *Didymoplexis micradenia* and *Epipogium roseum*.

Situated almost in the centre of Melanesia, the islands of Vanuatu have significant proportions of their orchid flora in common with Australia (14%) and New Caledonia (41%) in the southwest, New Guinea (42%), Bougainville and the Solomon Islands (62%), and the Malay archipelago (24%) in the northwest, and other eastern Pacific islands (53%) (see table 2). The flora of this island group thus exhibits a much greater intermingling of Palaeo-Oriental, Australian and Pacific elements than that of the Solomon Islands (Thorne and Cribb, 1984).

The proximity of New Caledonia with its strange flora has had a marked effect on the flora of the southern islands of Vanuatu. For example, *Megastylis gigas* and *Dipodium punctatum* var. *squamatum* are endemic to New Caledonia and the southern islands of Vanuatu.

Table 2: Number of Vanuatu orchid species recorded from other areas.

Area	No. of Vanuatu orchid species recorded from area	% Vanuatu orchid flora
Bougainville, the Solomon Islands and the Santa Cruz islands	93	62%
Pacific Islands (Fiji, Samoa, Horn Islands, Society Islands etc.)	80	53%
New Guinea	63	42%
New Caledonia	61	41%
Malay archipelago	36	24%
Australia	21	14%
Endemic	11	7%
New Zealand	3	2%

In this account 158 species of orchid have been recorded in 69 genera. This is probably an underestimate and the number is likely to increase as the islands become better known. The numbers of species recorded from the different islands are as follows (see table 3, map 3):

Table 3: Number of species recorded from the different islands in the Vanuatu archipelago (see map 3).

Northern region		Southern region	
Banks Islands	40	Efate	57
Espiritu Santo	95	Erromango	68
Malekula	23	Tanna	30
Ambae	45	Anatom	63
Maewo	5		
Pentecost	48		
Ambrym	7		
Epi	8		
Shepherd Islands	2		

These figures almost certainly reflect the amount of collecting done on each island as much as the species diversity of the islands, for there are few collections of orchids from the islands in the north-east and it would pay future expeditions to concentrate on this area. Pierre Cabalion (pers. comm.) has also noted that there are many rich areas of forest on north-west Espiritu Santo which have yet to be fully explored.

It is thought that orchids have reached the Pacific Islands by long distance wind dispersal of their seeds (Cribb, 1987). Cool temperatures, such as would be experienced if the seed were lifted by wind to high altitude, is known to prolong the viability of seed in the laboratory (Sanford, 1974), and the light seeds of orchids are adapted for long-distance wind-dispersal. The appearance of orchid species amongst the first colonisers of Krakatau is well documented (Doctors van Leeuwen, 1936). The predominant winds in the region, the south-east Trades, blow from south-east to north-west and are quite sufficient to take Asiatic orchids to the Pacific Islands, for example *Bulbophyllum longiflorum, Spiranthes sinensis, Liparis caespitosa* and *Epipogium roseum* are distributed from Africa to the Pacific Islands. The possibility of an orchid seed germinating and establishing itself is also dependent upon the presence of a suitable symbiotic fungal partner and we suggest that this requirement is probably the major limiting factor for orchids in the Pacific. The colonization of an island by an orchid species is thus a haphazard affair.

A further problem faces a colonizing orchid and this is the need for a pollinating agent. Some of the most widespread species have overcome this by being self-pollinating, for example, *Spiranthes sinensis, Liparis condylobulbon* and *Calanthe triplicata*. Most orchids are capable of vegetative reproduction and will survive long periods without sexual reproduction.

A Resource for Vanuatu

The forests of Vanuatu are a vital resource in the maintenance of the national livelihood, currently providing 7% of the total export earnings and they are also an essential source of wood supply for the Ni-Vanuatu population that live in the rural areas. The effective use of the forest resources requires planning based on information that includes accurate indentification of species. Orchids are a useful indicator of environmental species' richness and may therefore be a useful guide in selecting areas that need protection. There are also plans for a botanic garden to be set up at Tagabe, on Efate, in which orchids will feature prominently.

LIST OF NEW COMBINATIONS AND RECORDS

NEW COMBINATIONS

Peristylus maculiferus (C. Schweinf.) Renz & Vodonaivalu
Peristylus stenodontus (Reichb. f.) Renz & Vodonaivalu
Tuberolabium papuanum (Schltr.) J.J. Wood

NEW RECORDS

NEW GENUS RECORDS:

1. *Neuwiedia veratrifolia* Blume
2. *Erythrodes bicarinata* Schltr.
3. *Erythrodes oxyglossa* Schltr.
4. *Eurycentrum salomomense* Schltr.
5. *Cheirostylis montana* Blume
6. *Cryptostylis arachnites* (Blume) Hassk.
7. *Acanthephippium papuanum* Schltr.
8. *Epiblastus sciadanthus* (F. Muell.) Schltr.
9. *Octarrhena angraecoides* (Schltr.) Schltr.
10. *Cadetia quadrangularis* Cribb & B. Lewis
11. *Diplocaulobium ouhinnae* (Schltr.) Kraenzl.
12. *Pedilochilus hermonii* Cribb & B. Lewis
13. *Cleisostoma pacificum* Cribb & B. Lewis
14. *Microtatorchis schlechteri* Garay
15. *Pomatocalpa marsupiale* (Kraenzl.) J.J. Smith
16. *Eulophia nuda* Lindley

SPECIES RECORDS:

17. *Vrydagzynea argyrotaenia* Schltr.
18. *Corybas sp. nov.*
19. *Peristylus maculiferus* (C. Schweinf.) Renz & Vodonaivalu
20. *P. papuanus* (Kraenzl.) J.J. Smith
21. *P. wheatleyi* Cribb & B. Lewis
22. *Habenaria novaehiberniae* Schltr.
23. *Phaius robertsii* F. Muell.
24. *Liparis aaronii* Cribb & B. Lewis
25. *L. pullei* J.J. Smith
26. *Malaxis dryadum* (Schltr.) P.F. Hunt
27. *Agrostophyllum costatum* J.J. Smith
28. *A. graminifolium* Schltr.
29. *A. leucocephalum* Schltr.
30. *A. torricellense* Schltr.
31. *Appendicula bracteosa* Reichb. f.
32. *A. polystachya* (Schltr.) Schltr.
33. *Phreatia caulescens* Ames
34. *Glomera papuana* Rolfe

35. *Dendrobium spectabile* (Blume) Miq.
36. *D. aegle* Ridley
37. *D. delicatulum* Kraenzl.
38. *D. masarangense* Schltr.
39. *D. laevifolium* Stapf
40. *D. prosthecioglossum* Schltr.
41. *D. greenianum* Cribb & B. Lewis
42. *D. kietaense* Schltr.
43. *D. bilobum* Lindley
44. *Bulbophyllum stenophyllum* Schltr.
45. *B. microrhombos* Schltr.
46. *B. minutipetalum* Schltr.
47. *B. streptosepalum* Schltr.
48. *B. membranaceum* Teijsm. & Binnend.
49. *B. sp. nov.*
50. *B. polypodioides* Schltr.
51. *B. atrorubens* Schltr.
52. *Thrixspermum adenotrichum* Schltr.
53. *T. graeffei* Reichb. f.

CLASSIFICATION OF GENERA
Arranged according to Dressler (1981).

SUBFAMILY APOSTASIOIDEAE

1. *Neuwiedia*

SUBFAMILY SPIRANTHOIDEAE

Tribe Erythrodeae
2. *Corymborkis*
3. *Tropidia*
4. *Cystorchis*
5. *Erythrodes*
6. *Eurycentrum*
7. *Goodyera*
8. *Macodes*
9. *Moerenhoutia*
10. *Pristiglottis*
11. *Cheirostylis*
12. *Hetaeria*
13. *Vrydagzynea*
14. *Zeuxine*

Tribe Cranichideae
15. *Spiranthes*
16. *Cryptostylis*

SUBFAMILY ORCHIDOIDEAE

Tribe Diurideae
17. *Megastylis*
18. *Corybas*
19. *Microtis*

Tribe Orchideae
20. *Peristylus*
21. *Habenaria*

SUBFAMILY EPIDENDROIDEAE

Tribe Gastrodieae
22. *Nervilia*
23. *Didymoplexis*
24. *Gastrodia*

Tribe Epipogieae
25. *Epipogium*

Tribe Arethuseae
26. *Acanthephippium*
27. *Calanthe*
28. *Phaius*
29. *Spathoglottis*

Tribe Coelogyneae
30. *Coelogyne*
31. *Pholidota*

Tribe Malaxideae
32. *Liparis*
33. *Malaxis*
34. *Oberonia*

Tribe Epidendreae
35. *Eria*
36. *Trichotosia*
37. *Mediocalcar*
38. *Ceratostylis*
39. *Epiblastus*
40. *Agrostophyllum*
41. *Appendicula*
42. *Octarrhena*
43. *Phreatia*
44. *Aglossorhyncha*
45. *Earina*
46. *Glomera*
47. *Glossorhyncha*
48. *Cadetia*
49. *Dendrobium*
50. *Diplocaulobium*
51. *Flickingeria*
52. *Bulbophyllum*
53. *Pedilochilus*

SUBFAMILY VANDOIDEAE

Tribe Vandeae
54. *Gunnarella*
55. *Sarcanthopsis*
56. *Thrixspermum*
57. *Schoenorchis*
58. *Luisia*
59. *Cleisostoma*
60. *Microtatorchis*
61. *Pomatocalpa*
62. *Robiquetia*
63. *Taeniophyllum*
64. *Tuberolabium*

Tribe Cymbidieae
65. *Chrysoglossum*
66. *Eulophia*
67. *Geodorum*
68. *Oeceoclades*
69. *Dipodium*

KEY TO THE GENERA

1. Plant epiphytic 2
   Plant terrestrial 41
2. Plant covered in red-brown hairs ... ... **36. Trichotosia**
   Plant glabrous or finely pubescent, hairs never red-brown 3
3. Plant leafless; roots green, fasciculate, assuming the function of leaves
   ... ... **63. Taeniophyllum**
   Plant with leaves 4
4. Leaves terete 5
   Leaves not terete 8
5. Inflorescence terminal 6
   Inflorescence lateral 7
6. Stems 7–15 cm long; flowers sessile ... ... **38. Ceratostylis**
   Stems up to 2 m long; flowers in a lax raceme
   ... ... **48. Dendrobium** (*D. seemannii*)
7. Flowers yellow-green with apex of lip purple or rarely green; lip lacking
   a spur ... ...**58. Luisia**
   Flowers white; lip with globular spur ... ...**57. Schoenorchis**
8. Leaves bilaterally compressed 9
   Leaves dorso-ventrally flattened 10
9. Leaves more than 5 mm wide; inflorescence terminal ... ... **34. Oberonia**
   Leaves less than 3 mm wide; inflorescence lateral ... ... **42. Octarrhena**
10. Stems swollen, pseudobulbous 11
    Stems not pseudobulbous 25
11. Inflorescence lateral or from the base of the pseudobulb 12
    Inflorescence terminal 16
12. Inflorescence from the base of the pseudobulb 13
    Inflorescence lateral 15
13. Pseudobulbs with 1 or 2 leaves; flowers numerous, minute; pollinia 8
    ... ... **43. Phreatia** (in part)
    Pseudobulbs with 1 leaf; flowers solitary to many, small to large; pollinia 2
    or 4 14
14. Lip saccate ... ... **53. Pedilochilus**
    Lip not saccate, usually hinged to the column-foot
    ... ... **52. Bulbophyllum**
15. Pseudobulbs concealed by brownish tubular closely appressed sheaths, the
    upper portion bearing 5 or 6 crowded leaves; pollinia 8 ... ... **35. Eria**
    Pseudobulbs not as above; pollinia 4
    ... ... **49. Dendrobium** (in part only)
16. Plants small and creeping; flowers orange to cherry red with white or yellow
    tips; sepals connate in basal half; growing in rainforest above 500 m
    ... ... **37. Mediocalcar**
    Plants without the above combination of features 17
17. Stem branching, each branch of stem terminating in a pseudobulb, which
    bears a solitary, elliptic leaf; lip with apex of midlobe densely fringed
    ... ... **51. Flickingeria**
    Growth form not as above; lip apex not densely fringed 18
18. Stem consisting of 5–6 superposed pseudobulbs; flowers in clusters of up to
    8, red at base fading out to pink at tips ... ... **39. Epiblastus**
    Stem pseudobulbous; flowers not as above 19

19

38. Leaves present on young plant only; infloresence angled in cross section, may appear zig-zag ... ... **60. Microtatorchis**
    Plant with leaves; inflorescence straight 39
39. Leaves thick; inflorescence c. 1.5 cm long ... ... **64. Tuberolabium**
    Leaves thin; inflorescence 3.5 cm long or more ... ... **56. Thrixspermum**
40. Sepals c. 4 mm long; pollinia 4 ... ... **54. Gunnarella**
    Sepals less than 2 mm long; pollinia 8 ... ... **43. Phreatia** (in part only)
41. Plant saprophytic 42
    Plant with green leaves 45
42. Lip spurred ... ...**25. Epipogium**
    Lip not spurred 43
43. Sepals and petals free and spreading, pink or white with a yellowish-white lip spotted with pink ... ... **69. Dipodium**
    Sepals and petals fused basally to form a short tube, not opening widely 44
44. Sepals gibbous; flowers white with grey inside, lip with a yellow tip ... ... **24. Gastrodia**
    Sepals not gibbous; flowers pale brown, olive or pinkish with a yellowish-white lip spotted with pink ... ... **23. Didymoplexis**
45. Column showing partial fusion of the style and anthers; anthers 3 ... ... **1. Neuwiedia**
    Column showing complete fusion of the style and anthers; anther 1 46
46. Leaves rather fleshy, ovate-rotund, dark green with silver, greenish-yellow or pinkish venation ... ... **8. Macodes**
    Leaves not as above 47
47. Leaf broadly heart-shaped, solitary 48
    Leaves not with the above combination of features 49
48. Dorsal sepal large and helmet-shaped; lip with 2 spurs ... ... **18. Corybas**
    Dorsal sepal not as above; lip without a spur ... ... **22. Nervilia**
49. Leaf terete, solitary ... ... **19. Microtis**
    Leaves dorso-ventrally flattened, 1 or more 50
50. Inflorescence lateral or arising from the rhizome 51
    Inflorescence terminal 59
51. Inflorescence branched ... ... **2. Corymborkis**
    Inflorescence simple 52
52. Flowers large and urn-shaped, yellow lined and flushed with red ... ... **26. Acanthephippium**
    Flowers not as above 53
53. Peduncle strongly recurved just below flowers at maturity ... ... **67. Geodorum**
    Peduncle not recurved at maturity 54
54. Flowers without a spur 55
    Flowers spurred 56
55. Leaves and inflorescence arising separately from the rhizome; flowers non-resupinate; lip entire ... ... **16. Cryptostylis**
    Stem leafy; inflorescence from a basal leaf axil; flowers resupinate; lip trilobed with a callus at the base between the lateral lobes ... ... **29. Spathoglottis**
56. Lateral lobes of lip inrolled to form a tube which encompasses the column ... ... **28. Phaius**
    Lip not as above 57
57. Flowers turning blue when damaged; spur clavate to cylindrical, 0.5–2.5 cm long; pollinia 8 ... ... **27. Calanthe**

Flowers not turning blue when damaged; spur small and globular, c. 3 mm long; pollinia 2 58

58. Lip entire; sepals 2.2 cm long or more ... **66. Eulophia**
    Lip 4-lobed; sepals 1 cm long ... **68. Oeceoclades**

59. Lip with median constriction, marging of mesochile sinuous
    ... **10. Pristiglottis**
    Lip not as above 60

60. Flowers numerous, arranged spirally ... **15. Spiranthes**
    Flowers one to many but never spirally arranged 61

61. Lip with a spur 62
    Lip without a spur 68

62. Lip entire 63
    Lip 3-lobed 65

63. Rhachis and ovary pubescent 64
    Rhachis and ovary glabrous 65

64. Spur cylindric-conical, c. 4 mm long, with 2 sessile glands at the apex
    ... **6. Eurycentrum**
    Spur bilobed, 2–3 mm long, without glands ... **5. Erythrodes**

65. Glands stalked ... **13. Vrydagzynea**
    Glands not stalked ... **4. Cystorchis**

66. Pseudobulb with 1 leaf or an erect inflorescence
    ... **65. Chyrsoglossum**
    Stem with leaves scattered or in an apical rosette 67

67. Stigma sessile or very shortly stalked, entire or partly divided
    ... **20. Peristylus**
    Stigmas lobes distinctly stalked, 2 ... **21. Habenaria**

68. Leaves narrowly lanceolate, grouped at base of stem; inflorescence sheathed in cataphylls; dorsal sepal hooded, 2.5–4 cm long; flowers white
    ... **17. Megastylis**
    Leaves lanceolate to ovate, distichous; dorsal sepal may be hooded but less than 1.5 cm long 69

69. Ovaries and sepals glabrous 70
    Ovaries ± sepals pubescent 72

70. Leaves lanceolate; petioles not sheathing; margins of petal and lip sinuous
    ... **3. Tropidia**
    Leaves ovate; petioles sheathing; margins of petals and lip not sinuous ... 71

71. Column very short, embraced by base of lip; lip flat, never ovate
    ... **33. Malaxis**
    Column long, not embraced by base of lip; lip porrect to recurved, never flat, broadly ovate ... **32. Liparis** (in part)

72. Lip apex widened into a transversely oblong blade 73
    Lip apex not widened into a blade 74

73. Flowers with sepals fused for about half their length to form a tube which is swollen at the base; lip with a saccate base containing a few papillae, extended above into a deeply bilobed lamina, with toothed lobes
    ... **11. Cheirostylis**
    Flowers with sepals not fused; lip with a saccate base containing 2 glands, lamina not toothed ... **14. Zeuxine**

74. Lip without papillae or glands but may be hairy ... **7. Goodyera**
    Lip with basal papillae or glands 75

75. Lip uppermost, with basal glands; column not winged ... **12. Hetaeria**
    Lip at the bottom, with basal papillae; column winged **9. Moerenhoutia**

SUE
WICKISON

DESCRIPTIONS OF THE GENERA AND SPECIES

1. **NEUWIEDIA** Blume

Terrestrial, with aerial roots and with a short rhizome. *Stems* erect. *Leaves* lanceolate to linear, acuminate, plicate, upper ones hairy. *Inflorescence* terminal, erect, unbranched, many-flowered, hairy. *Flowers* nodding, not opening widely; sepals and petals subsimilar; lip similar to the sepals and petals, with a central longitudinal but obscure callus; column showing partial fusion of style and anthers; anthers 3. *Fruit* a thin-walled capsule.

A small genus of about 8 species from tropical S.E. Asia to the Pacific Islands. A new genus record for Vanuatu, with a single species recorded.

N. veratrifolia *Blume* in van d. Hoeven & de Vriese, Tijd. Nat. Gesch. Phys. 1: 140 (1834). Type: Java, *Blume* s.n. (lectotype L!).
For full synonymy see de Vogel (1969).

Plant 70–120 cm tall, growing in small colonies. *Rhizome* 4 mm diameter. *Leaves* erect, lanceolate, acuminate, up to 78 cm long, 6.7 cm wide. *Flowers* yellow, fragrant; sepals narrowly elliptic-lanceolate, shortly apiculate, 1.4–1.6 cm long; petals obovate, apiculate, with a central fleshy ridge, 1.6 cm long; lip obovate, apiculate, with a central fleshy ridge, 1.5 cm long, 0.7 cm wide; column with three anther filaments united at base only to style, 1.2 cm long. *Fruit* triangular in cross-section, 1–1.4 cm long, pale green. (See fig. 1).

DISTRIBUTION: Banks Islands (Vanua Lava). Also in Sumatra and the Malay peninsula to the Philippines, New Guinea and the Solomon Islands.
HABITAT: Well drained primary forest in deep leaf litter overlaying calcareous soil and limestone outcrops, 200–370 m.
COLLECTION: *Wheatley* 344 (K, PVNH).

2. **CORYMBORKIS** Thouars

Terrestrial, with subterranean rhizomes. *Stems* erect, slender but woody, unbranched. *Leaves* in upper part of stem, plicate, rather broad, thin but tough. *Inflorescences* lateral, paniculate. *Flowers* with long, slender, spathulate sepals and petals; lip about the same length as sepals and petals, slender, except for an apical, decurved lamina, disc with 2 longitudinal keels; column straight, long and slender, dilated at the apex, rostellum erect; pollinia 2.

A small genus of about 5 species from the Old World Tropics. A single species in Vanuatu.

C. veratrifolia (*Reinw.*) *Blume* in Orch. Arch. Ind.: 125 (1855). Type: Java, *Lobb* 162 (neotype K!).
For full synonymy see Rasmussen (1977).

Fig. 1. *Neuwiedia veratrifolia.* **A**, habit × ⅛; **B**, leaf ⅔; **C**, fruit × ⅔; **D** inflorescence × ⅔; **E**, dorsal sepal, anthers and column × 2; **F**, lateral sepal × 2; **G**, petal × 2; **H**, lip × 2. **A** & **C** drawn from *Fa'arodo et al* BSIP 12256; **B** & **D** from *Streimann* NGF 24478; **E–H** from *Dennis* s.n. (Kew spirit no. 45741). All drawn by Sue Wickison.

Stem up to 2 m high, erect. *Leaves* 6–15, ovate or obovate, acute to acuminate, 18–35 cm long, 5–10 cm wide, ribbed and fluted, dark green. *Inflorescences* 1–4, corymbose-paniculate, consisting of up to 60 flowers. *Flowers* green and white, scented; sepals and petals linear-lanceolate, expanded at apices, irregularly recurved, 2–3 cm long; lip about the same length as the sepals, basal portion narrow, embracing column, with 2 barely discernible, long, raised lines which diverge on the lamina and terminate in small calli, apical lamina round, reflexed, with irregularly undulate margins; column 9–10 mm long.

DISTRIBUTION: Banks Islands (Gaua, Vanua Lava), Efate, Epi, Erromango, Espiritu Santo, Maewo, Malekula, Pentecost, the Shepherd Islands (Tongoa) and Tanna. Widely distributed from S.E. Asia and the Malay archipelago to New Guinea, the Solomon Islands, Samoa and Australia.

VERNACULAR NAME: Nëré vudhvékar.

HABITAT: Rain forest, in *Kleinhovia-Dendrocnide* forest and in disturbed *Castanospermum* forest, on alluvial flats and coral limestone, 30–500 m.

COLLECTIONS: *Baker* 3 (BM); *Bernardi* 13305 (K, P, G) & 13335 (P, G); *Bourdy* 286 & 927 (K, P, PVNH), 1113 (P, PVNH); *Bregulla* 21 (K); *Cabalion* 1124 & 1156 (PVNH), 1532 (P, PVNH) & 3124 (P); *Hallé* in RSNH 6312 (K, P, PVNH); *Kajewski* 146 & 447 (K); *Morat* 5180 & 5226 (P); *Morrison* in RBG Kew 65 & 77 (K); *Raynal* in RSNH 16415 (K, P, PVNH); *de la Rüe* s.n. (P); *Sam* 347 (P, PVNH); *Suprin* 355 (P); *Veillon* 2392 (P); *Chew Wee-Lek* in RSNH 271 & 383 (K, P, PVNH), 332A (K, P); *Wheatley* 199 (K, PVNH).

3. **TROPIDIA** Lindley

Terrestrial. Stems erect. *Leaves* distichous, tough, plicate. *Inflorescence* terminal, simple, short and dense. *Flowers* not opening widely; sepals and petals free; lateral sepals larger than the dorsal sepal; lip not lobed, concave; column short, rostellum long and erect; pollinia 2.

A pantropical genus of about 20 species. A single species in Vanuatu.

T. viridifusca *Kraenzl.* in Viertelj. Nat. Ges. Zur. 74: 71 (1929). Type: New Caledonia, *Daeniker* 2720 (holotype Z).

Stem 25–40 cm high. *Leaves* up to 7, towards apex of stem, distichous, lanceolate, 10–20 cm long, 1–2 cm wide. *Inflorescence* terminal, short, c. 1 cm long; peduncle sheathed in imbricate bracts. *Flowers* green, with brown or purple patches, scent unpleasent; pedicels c. 1 cm long; dorsal sepal ovate, 10–12 mm long; lateral sepals broader than dorsal sepal, 10 mm long; petals lanceolate, 8.5–9.5 mm long, with wavy margins; lip ovate, acute, 9 mm long, 3 mm wide, saccate at base, fleshy, with sinuous margins; disc with 2 fleshy calli.

DISTRIBUTION: Anatom. Also in New Caledonia and on Norfolk Island.

HABITAT: Forest.

COLLECTION: *Morrison* in RBG Kew 104 (K) (in fruit).

R. Hoogland (pers. comm.) has noted that on Norfolk Island this species is fly-pollinated.

4. **CYSTORCHIS** Blume

Terrestrial. Stem leafy, arising from creeping base. *Leaves* ovate; petioles sheathing. *Inflorescence* long. *Flowers* with dorsal sepal and petals adnate to form a hood; lateral sepals enclosing base of lip; lip partly fleshy, edges infolded, spur short, with a swollen vescicle containing a gland at the base on either side; column short; pollinia 2.

A small genus of about 10 species from the Malay Archipelago to the Pacific Islands. A single species in Vanuatu.

C. variegata (Miq.) *Blume*, Fl. Java: 74, tab. 24, fig. 3 (1858).
Hetaeria variegata Miq., Fl. Ind. Bat. 3: 726 (1855). Type: Java, *Blume* 902 (holotype BO; isotype L).

Plant 15–25 cm tall. *Leaves* c. 6, near the base, elliptic, acute, 7 cm long, 2.8 cm wide, asymmetric, light green with a darker network of veins; petiole c. 2 cm long. *Inflorescence* pubescent, 10–15 cm long; rhachis c. 3 cm long. *Flowers* 7–12, not opening widely; sepals pinkish–brown, becoming yellowish green at base, petals white, lip white and orange; sepals c. 6 mm long; lateral sepals concave at base, enclosing the base of the spur; lip straight, c. 6 mm long, fleshy with a central groove, with margins of apical half inflexed to form a tube, spur short and slightly decurved, projecting between the lateral sepals, with a swollen vesicle containing a gland on each side.

DISTRIBUTION: Efate. Also in the Malay peninsula and archipelago.
HABITAT: No information from Vanuatu. In Java it occurs in lowland rain forest.
COLLECTION: No specimens seen, (Guillaumin, 1929 & 1948).

Guillaumin cites a specimen, *Levat* s.n., from Efate as being *Cystorchis variegata*. Unfortunately we have not been able to study this specimen and the description is from collections from Java.

5. **ERYTHRODES** Blume

Terrestrial. Leaves distichous, ovate, with unequal halves, green; petioles sheathing. *Inflorescence* terminal, loosely racemose; rhachis and ovary pubescent. *Flowers* with dorsal sepal and petals forming a hood; lip spurred, concave, with a reflexed apex, spur 2-lobed, projecting between the lateral sepals; column short; pollinia 2.

A genus of about 60 species distributed throughout the tropics, except Africa. A new genus record for Vanuatu, two species being recorded.

Lip obovate, rounded and deflexed at apex ... ...   **1. E. bicarinata**
Lip ovate, acute   ... ...   **2. E. oxyglossa**

1. E. bicarinata *Schltr.* in Fedde, Rep. Sp. Nov. Beih. 1: 61 (1912). Type: New Guinea, *Schlechter* 20012 (holotype B).

Stem erect, up to 8 cm long. *Leaves* 4–6, obliquely ovate, acute, 2.5–5 cm long,

1-2 cm wide, satiny grey-green with darker venation, slenderly petiolate. *Inflorescence* laxly 10-15-flowered, 10-18 cm long, pubescent; bracts lanceolate, slenderly acuminate, 3-8 mm long. *Flowers* pubescent, with light russet-brown sepals, white petals and a white lip; ovary 7-8 mm long; dorsal sepal oblong-elliptic, rounded at apex, 3 mm long; lateral sepals obliquely ovate, obtuse, 3.5-4 mm long; petals obovate, obtuse, 3-3.5 mm long; lip obovate, rounded and deflexed at apex, 3 mm long, 2 mm wide, spur broad, straight, bilobed at apex, 2.5-3 mm long; column 3 mm long.

DISTRIBUTION: Banks Islands (Vanua Lava). Also in New Guinea.
HABITAT: In deep shade in forest, 390-410 m.
COLLECTION: *Wheatley* 347 (K, PVNH).

2. E. oxyglossa *Schltr.* in Engler, Bot. Jahrb. 39: 53 (1906). Type: New Caledonia, *Schlechter* 15749 (holotype B).

Plant up to 30 cm high. *Leaves* 4-6, distichous, obliquely ovate, 5-8 cm long, 1.5-2 cm wide; petioles 2-2.5 cm long, sheathing. *Inflorescence* terminal, up to 16 cm long; raceme and pedicels pubescent. *Flowers* 4-8, brown to dull pink; sepals pubescent outside; dorsal sepal ovate, 6.5 mm long; lateral sepals obliquely ovate, 8 mm long; petals rhombic, 6.5 mm long; lip ovate, acute, 8.5 mm long, 3 mm wide, spur bilobed, c. 2 mm long.

DISTRIBUTION: Efate. Also in New Caledonia, the Horn Islands and Fiji.
HABITAT: Lower montane forest, 480 m.
COLLECTION: *Sam* CSV 76 (NOU, P).

6. **EURYCENTRUM** Schlechter

Terrestrial. Rhizome fleshy, creeping. *Stems* fleshy, erect and leafy. *Leaves* several, ovate, very dark green above, purple below. *Inflorescence* terminal, laxly many-flowered, pubescent. *Flowers* small, sessile, resupinate, often pubescent on outer surface; dorsal sepal forming a hood over the column; lateral sepals free, spreading; petals smaller than sepals, adnate to the dorsal sepal; lip bipartite or obscurely trilobed, somewhat pandurate, prominently spurred at the base, basal part concave, apical part small and ovate to transversely oblong, spur with a broad mouth and two sessile glands at the apex inside; column porrect, short, rostellum truncate; pollinia 4.

A small genus of about 5 species from New Guinea and the Pacific Islands. A new genus record for Vanuatu, with a single species being recorded.

E. salomonense *Schltr.* in K. Schum. & Laut. , Nachtr. Fl. Deutsch. Sudsee: 90, t. 5, fig. B (1905). Type: Solomon Islands, *Guppy* 78 (holotype B; isotypes K!, MEL).

Plant 20-35 cm tall; rooting at the nodes of the rhizome. *Leaves* 4-7, ovate, acute, 4-7.5 cm long, 1.5-2.8 cm wide, satiny and almost black to very dark green and with a pink central vein above, purple beneath. *Inflorescence* 18-23 cm long, laxly many-flowered; peduncle pubescent, slender; bracts lanceolate, setose, 4-9 mm long, red. *Flowers* off white with red sepals; ovary 4-8 mm long, pubescent; dorsal sepal concave, ovate, obtuse to bluntly apiculate, 2-3 mm long; lateral sepals obliquely ovate, acuminate, 2-2.5 mm long; lip concave, 2 mm long, 1.5 mm wide, obscurely trilobed, lateral lobes rounded, midlobe tiny, shortly clawed, ovate, obtuse, spur compressed, cylindric-conical, 4 mm long; column 1 mm long.

DISTRIBUTION: Banks Islands (Vanua Lava). Also in the Solomon Islands.
HABITAT: Primary forest, up to 400 m.
COLLECTIONS: *Cheesman* s.n. (K); *Wheatley* 396 (K, PVNH).

7. **GOODYERA** R. Brown

Terrestrial or rarely epiphytic. *Leaves* ovate, often asymmetric; petioles sheathing stem. *Stem*, peduncle and ovaries finely pubescent. *Inflorescence* terminal. *Flowers* 1 to many, usually not opening widely, pale green, light pink to dull brown; dorsal sepal and petals hooded; lip concave, glabrous or papillose; column long with a long erect rostellum; pollinia 2.
A cosmopolitan genus of about 50 species. Three species in Vanuatu, *Goodyera rubicunda* var. *rubicunda* being a new record.

1. Outer surface of sepals glabrous ... ... **3. G. viridiflora**
   Outer surface of sepals pubescent 2
2. Leaves with pale venation, ovate; bracts shorter than or as long as ovary; lip lanceolate, glabrous ... ... **2. G. subregularis**
   Leaves without pale venation, obliquely ovate; bracts distinctly longer than the ovary; lip ovate, saccate, papillose at base 3
3. Anther single ... ... **1. G. rubicunda** var. **rubicunda**
   Anther divided into 3 ... ... **1. G. rubicunda** var. **triandra**

1. G. rubicunda (*Blume*) *Lindley* in Bot. Reg. 25: 61, misc. n. 92 (1839). Type: Java, *Blume* s.n. (isotype P!).
For full synonymy see Hallé (1977).

Stem up to 17 cm high. *Leaves* 5–10, obliquely ovate, 9–13 cm long, 3.2–6 cm wide; petiole 3–4 cm long. *Inflorescence* 20–30 cm long. *Flowers* 10–20, not opening widely, dark brown to orange to pale pink, white to yellow inside; exterior of sepals pubescent; sepals and petals c. 8 mm long; sepals ovate; petals spathulate; lip ovate, saccate, papillose at base, c. 5 mm long, 3.5 mm wide, tongue-shaped at deflexed apex, pale yellow; column c. 3 mm long, pale yellow.

var. **rubicunda**

Anther undivided.

DISTRIBUTION: Pentecost. Also in the Malay archipelago, Indonesia, New Guinea and Fiji.
HABITAT: Rain forest in deep shade, 480 m.
COLLECTION: *Wheatley* 116 (K, PVNH).

var. **triandra** (*Schltr.*) *N. Hallé*, Fl. Nouv. Caled. 8: 532 (1977).
Goodyera triandra Schltr. in Bull. Herb. Boiss., ser 2. 6: 298 (1906). Type: Vanuatu, Efate, *Morrison* s.n. (holotype B).

Goodyera anomala Schltr. in Fedde, Rep. Sp. Nov. 9: 86 (1910). Type: Samoa, *Vaupel* 405 (holotype B).

Anther is divided into three.

DISTRIBUTION: Efate and Malekula. Also in New Caledonia, Fiji and Samoa.
HABITAT: Rain forest.
COLLECTIONS: *Sam* CSV 76 (PVNH); *Hallé* in RSNH 6387, 6392 & 6417 (P); *Morrison* in RBG Kew 143 & 144 (K).

2. G. subregularis *(Reichb. f.) Schltr.* in Engler, Bot. Jahrb. 36: 25 (1906).
Georchis subregularis Reichb. f. in Linnaea 41: 61 (1877). Type: New Caledonia, *Vieillard* 1312 (holotype P!).

Stem up to 7 cm high. *Leaves* 4–5, ovate, (2)3–6 cm long, (1.5)2–3.5 cm wide, dark green to velvety brown with pale or orange veins; petiole 1–1.5 cm long. *Inflorescence* 7–10 cm long. *Flowers* 1–7, dull brown, with exterior of the sepals pubescent; sepals and petals lanceolate, c. 10 mm long; lip lanceolate, c. 9 mm long; column c. 8 mm long.

DISTRIBUTION: Anatom, Banks Islands (Vanua Lava), Efate, Erromango and Pentecost. Also in the Santa Cruz Islands and New Caledonia.
HABITAT: Dense rain forest, 290–780 m.
COLLECTIONS: *Cabalion* 1931 (P) & 2353 (PVNH); *Green* in RSNH 1166 (K, P); *Kajewski* 946 (K); *Morrison* in RBG Kew 21, 24, 25 & 26 (K); *de la Rüe* s.n. (P); *Wheatley* 357 (K, PVNH).

3. G. viridiflora *(Blume) Blume*, Fl. Java n. ser. 1: 34 (1858).
Neottia viridiflora Blume, Bijdr. Fl. Ned. Ind.: 408 (1825). Type: Java, *Blume* s.n. (holotype L; isotype P).
Georchis cordata Lindley, Gen. Sp. Orch. Pl.: 496 (1830). Type: Ceylon, *Macrae* s.n. (holotype K!).
Physurus viridiflorus (Blume) Lindley in Journ. Linn. Soc. 1: 180 (1857).
Goodyera cordata (Lindley) Hook. f., Fl. Brit. Ind. 5: 114 (1890).
Orchiodes viridiflorum (Blume) O. Kuntze, Rev. Gen. Pl. 2: 675 (1891).
Epipactis viridiflora (Blume) Ames, Orch. 2: 61 (1908).
Goodyera finetiana Kraenzl. in Not. Syst. 4: 138 (1928). Types: New Caledonia, *Le Rat* 899 & *Pancher* 643 (syntypes P!).

Stem up to 10 cm high, pale to silver green. *Leaves* 4–5, obliquely ovate, 5.5–7 cm long, 2.5–3.5 cm wide, mid green with a slightly darker venation; petiole 2–4 cm long. *Inflorescence* 14–20 cm long. *Flowers* 7–13; pale green turning orange with age; exterior of sepals glabrous; sepals and petals c. 9 mm long; sepals ovate; petals rhomboid; lip oblong, 5.5 mm long, 3.5 mm wide, translucent white, saccate and papillose at base, acute and deflexed at apex; column c. 7 mm long. (See plate 1e).

DISTRIBUTION: Ambae, Anatom and Espiritu Santo. Widely distributed from Asia to the Malay archipelago, New Guinea, the Solomon Islands, New Caledonia and Australia.
HABITAT: Montane forest, 600–1600 m.
COLLECTIONS: *Cribb & Wheatley* 19 (K, PVNH); *Green* in RSNH 1161 (K, P); *Raynal* in RSNH 16384 (P); *Raynal & Schmid* in RSNH 16126 (P); *Wheatley* 40 (K, PVNH).

Goodyera brachiorrhynchos Schltr. from New Guinea is similar and may be conspecific.

8. MACODES Blume

Terrestrial. Leaves rather fleshy, broad, with coloured veins. *Flowers* rather small, with lip uppermost; lateral sepals enclosing base of lip; lip trilobed, saccate at base, containing 2 glands, midlobe with a narrow base and a short spreading blade; pollinia 2.

A genus of about 10 species from the Malay archipelago to the Pacific Islands. A single species in Vanuatu.

M. sanderiana (*Kraenzl.*) *Rolfe* in Kew Bull. 1896: 47 (1896).
Anoectochilus sanderianus Kraenzl. in Gard. Chron. ser. 3, 18: 484 (1895). Type: Sunda Islands, *Sander* s.n. (holotype B).
Anoectochilus spec. ex. aff. *A. roxburghii* Kraenzl. in K. Schum. & Laut., Nachtr. Fl. Deutsch. Sudsee: 240 (1900).

Stem 2–4 cm high. *Leaves* ovate-rotund, 4.5–8 cm long, 3–5 cm wide, dark olive green with silver, greenish-yellow or pinkish veins. *Inflorescence* with peduncle 20–30 cm long, pinkish fawn; raceme c. 7–15 cm long; peduncle and pedicel pubescent. *Flowers* 10–20, pinkish white, yellow or pale green lightly suffused with brown; pedicels c. 10 mm long, hairy; sepals hairy, concave, 6–7 mm long; dorsal sepal ovate; lateral sepals oblique-ovate; petals linear, falcate; lip trilobed, c. 6 mm long, 4 mm wide, saccate at base, containing 2 glands, lateral lobes short, rounded, midlobe with a narrow base and short spreading blade; column c. 4 mm long. (See plate 1b & d).

DISTRIBUTION: Anatom and Banks Islands (Vanua Lava). Also in the Malay archipelago, Sumatra and New Guinea.
HABITAT: Terrestrial in shady positions; up to 550 m.
COLLECTIONS: *Cabalion* 1985 (PVNH) (sterile); *Morrison* in RBG Kew 71 & 72 (K) (sterile); *Wheatley* 374 (K, PVNH) (sterile).

All collections of this species from Vanuatu, which we have examined, are sterile and therefore the description of the inflorescence is from New Guinea collections.

9. MOERENHOUTIA Blume

Terrestrial. Leaves ovate; petioles sheathing. *Inflorescence* terminal, racemose; peduncle and ovaries finely pubescent. *Flowers* many, not opening widely; petals spathulate; lip ovate, concave, constricted in middle, with basal papillae; column with 2 long, apical appendages; pollinia 2.

A genus of about 12 species in the Pacific Islands. A single species in Vanuatu.

M. grandiflora (Schltr.) Schltr. in Engler, Bot. Jahrb. 56: 450 (1921).
Goodyera grandiflora Schltr. in Engler, Bot. Jahrb. 39: 57 (1906). Type: New Caledonia, *Schlechter* 15750 (holotype B).

Platylepis morrisonii Schltr. in Fedde, Rep. Sp. Nov. 9: 161 (1911); **synon. nov.** Type: Vanuatu, Anatom, *Morrison* in RBG Kew 139 (holotype K!).
Moerenhoutia morrisonii (Schltr.) Schltr. in Engler, Bot. Jahrb. 56: 450 (1921); **synon. nov.**
Goodyera vieillardii Kraenzl. in Not. Syst. 4: 139 (1928). Type: New Caledonia, *Vieillard* 3277 (holotype P!).

Stem leafy, arising from a creeping base, up to 30 cm tall. *Leaves* 4–7, elliptic to ovate, 10–14 cm long, 4.5–5 cm wide; petioles sheathing, up to 5 cm long. *Inflorescence* terminal, up to 23 cm long; peduncle and ovaries pubescent; bracts as long as the ovary, pubescent. *Flowers* 10–20, white; pedicels 13 mm long; sepals glabrous, ovate, c. 11 mm long; petals spathulate, c. 10 mm long; lip ovate, concave, constricted in the middle, c. 9 mm long, 4 mm wide, with incurved margins and 2 areas of papillae at base; column c. 7 mm long, including elongated apical appendages.

DISTRIBUTION: Anatom and Efate. Also in New Caledonia.
HABITAT: Rain forest, 10–500 m.
COLLECTIONS: *Bernardi* 12962 (K, G); *Cabalion* 1280 & 2853 (P); *Morrison* in RBG Kew 139, 140, 141 & 142 (K).

Moerenhoutia heteromorpha (Reichb. f.) Drake from Samoa is similar and may prove to be conspecific.

10. **PRISTIGLOTTIS** Cretzoiu & J.J. Smith

Terrestrial or rarely epiphytic. *Stems* leafy, arising from a creeping rhizome. *Leaves* distichous, ovate; petioles sheathing. *Inflorescence* terminal, short, of few flowers. *Flowers* with sepals and petals not spreading; lateral sepals enclosing saccate base of lip; lip narrowed to a long claw with sinuous margins, widened to a blade at apex, saccate at base with 2 glands; column with a long rostellum; pollinia 2.
A small genus of about 3 species from New Guinea to the Pacific islands. A single species in Vanuatu.

P. montana (*Schltr.*) *Cretz. & J.J. Smith* in Acta Fauna Fl. Univ. Bucur., ser. 2, Bot. 1, 14: 4 (1934).
Anoectochilus montanus Schltr. in Engler, Bot. Jahrb. 39: 55 (1906). Type: New Caledonia, *Schlechter* 14933 (holotype B).
Cystopus aneytyumensis Schltr. in Fedde, Rep. Sp. Nov. 9: 282 (1911). Type: Vanuatu, Anatom, *Morrison* in RBG Kew 67 (holotype B; isotype K!).
Cystopus montanus (Schltr.) Schltr. in Engler, Bot. Jahrb. 45: 393 (1911).
Pristiglottis aneytyumensis (Schltr.) Cretz. & J.J. Smith in Acta Fauna Fl. Univ. Bucur., ser. 2, Bot. 1, 14: 4 (1934).

Plant up to 20 cm high. *Leaves* 3–8, ovate, 2–6 cm long, 1.5–2.4 cm wide, dark silky-velvety green, with mid-vein and marginal lateral veins pale emerald green, margins undulate. *Flowers* 2–4, white but base of lip may be green; pedicels c. 1 cm long; pedicels and outside of sepals pubescent; sepals 1–1.5 cm long; lip 11–17 mm long, with apical blade c. 4.5 mm wide. (See plate 1f).

DISTRIBUTION: Ambae, Anatom, Erromango, Espiritu Santo and Tanna. Also in New Caledonia and Samoa.

HABITAT: Rain forest, 200–1600 m.

COLLECTIONS: *Bernardi* 12980, 13000, 13205, 13317, 13361 (G), 13094 (K, G, P); *Cabalion* 720 (NOU), 1972 (PVNH), 2149 (P) & 2992 (P, PVNH); *Cribb & Wheatley* 98 (K, PVNH); *Green* in RSNH 1123 & 1246 (K); *Morat* 6057 (P); *Morrison* in RBG Kew 66, 67 & 69 (K); *Raynal* in RSNH 16150, 16182 & 16383 (P); *Veillon* 3994 & 4039 (P); *Wheatley* 34 (K, PVNH).

11. **CHEIROSTYLIS** Blume

Small *terrestrial* herbs. *Stem* leafy, arising from a fleshy creeping base. *Inflorescence* terminal, erect, with few flowers. *Flowers* with sepals fused for about half their length to form a tube which is swollen at the base; lip with a saccate base containing a few papillae, extended above into a deeply bilobed lamina, with toothed lobes; column short, thickened; pollinia 2.

A genus of about 20 species from tropical Africa and Asia to the Pacific Islands. A new genus record for Vanuatu, a single species being recorded.

C. montana *Blume*, Bijdr. Fl. Ned. Ind.: 413 (1825). Type: Java, *Blume* s.n. (holotype L).

Plant 14–30 cm tall. *Leaves* 6–9, ovate, 2–7 cm long, 1.3–3 cm wide, dark green; petiole 0.5–1.5 cm long. *Inflorescence* with a 10–16 cm long peduncle sheathed by about 4 sterile bracts; rhachis up to 4 cm long; peduncle, rhachis, bracts, ovary and outer sepals slightly hairy; bracts about as long as ovary at flowering. *Flowers* 2–8; sepals green, petals and lip white; sepals c. 5 mm long; lip 3–4 mm long when flattened.

DISTRIBUTION: Espiritu Santo. Widely distributed from Thailand and the Malay peninsula to New Caledonia.

HABITAT: Submontane and montane ridge-top forest, 400–1400 m.

COLLECTIONS: *Cribb & Wheatley* 49 & 104 (K, PVNH); *Raynal* in RSNH 16391 (P).

12. **HETAERIA** Blume

Terrestrial. Leaves broad, asymmetric, petiolate with the petioles sheathing. *Inflorescence* apical. *Flowers* non-resupinate, numerous, not opening widely; lip uppermost; dorsal sepal and petals forming a hood; lateral sepals enclosing the saccate base of the lip; lip uppermost in flower, concave, narrowed towards the apex, glandular within at the base; column short with large lateral wings; pollinia 2.

A genus of about 20 species in the Old World Tropics. A single species in Vanuatu.

H. oblongifolia Blume, Bijdr. Fl. Ned. Ind.: 410 (1825). Type: Java, *Blume* s.n. (holotype L).
Aetheria oblongifolia (Blume) Lindley, Gen. Sp. Orch. Pl. 491 (1840).

Rhamphidia discoidea Reichb. f. in Linnaea 41: 59 (1877); **synon. nov.** Type: New
  Caledonia, *Vieillard* 1311 (holotype P!).
Hetaeria forcipata Reichb. f. in Linnaea 41: 62 (1877). Type: Fiji, *Roezl* s.n.
  (holotype W).
Hetaeria samoensis Rolfe in Kew Bull. 1898: 199 (1898). Type: Samoa, *Walter* s.n.
  (holotype CAM, isotype K!).
Goodyera discoidea (Reichb. f.) Schltr. in Engler, Bot. Jahrb. 39: 57 (1906); **synon.
  nov.**
Hetaeria similis Schltr. in Fedde, Rep. Sp. Nov. 9: 88 (1910). Type: Samoa, *Vaupel*
  657 (holotype K!).
Hetaeria discoidea (Reichb. f.) Schltr. in Fedde, Rep. Sp. Nov. 9: 89 (1911); **synon.
  nov.**
Hetaeria similis Schltr. in Engler, Bot. Jahrb. 56: 453 (1921). Type: Palau Islands,
  Raymundus s.n. (holotype B).

Stem up to 30 cm long. *Leaves* 3–9, of which 3–4 are dead at flowering time and
the remainder are in a rosette, obliquely ovate, 7–9 cm long, 3–4 cm wide;
petioles c. 2 cm long. *Inflorescence* up to 30 cm long. *Flowers* 15–40, yellow, or white
with a yellow lip; peduncle, rhachis, ovaries and sepals hirsute; dorsal sepal
oblong to ovate, c. 3.6 mm long; lateral sepals ovate, c. 3.5 mm long; petals linear,
c. 3 mm long; lip ovate, c. 3.5 mm long, 2.5 mm wide, saccate, with 3–6 glands on
either side at the base, margins inrolled.

DISTRIBUTION: Erromango, Espiritu Santo and Malekula. Widely distributed
from the Philippines and Indonesia through New Guinea and the Palau Islands,
to the Solomon Islands, the Santa Cruz Islands, New Caledonia, Fiji, Samoa and
Australia.
HABITAT: Rain forest, in shade, 50–520 m.
COLLECTIONS: *Hallé* in RSNH 6390 (K, P, PVNH) & 16380 (PVNH);
MacGillivray 1 (P); *Raynal* in RSNH 16424 (P).

Hetaeria erimae Schltr. from New Guinea is similar and may prove to be
conspecific.

13. **VRYDAGZYNEA** Blume

Terrestrial small, erect herbs. *Stems* not pseudobulbous. *Leaves* 4–9, often
oblique, shortly petiolate at base, distichous on stem. *Inflorescence* terminal,
usually short and dense but elongating after pollination. *Flowers* few to many,
white or cream; sepals and petals not spreading; spur projecting between lateral
sepals, long, with 2 stalked glands inside; column very short; pollinia 2.
  A genus of about 40 species from S.E. Asia to Australia and the Pacific Islands.
Three species in Vanuatu, *Vrydagzynea argyrotaenia* being a new record.

1. Leaves variegated, dark green checkered with light green; flowers 1–4
   ... ... **1. V. argyrotaenia**
   Leaves not variegated; flowers 10–20 2
2. Leaves lanceolate, 2.7–5.5 cm long, 0.9–1.7 cm wide; spur apex minutely
   bilobed ... ... **3. V. whitmeei**
   Leaves obliquely ovate, 4–10 cm long, 2–3.5 cm wide; spur apex entire
   ... ... **2. V. salomonensis**

1. V. argyrotaenia *Schltr.* in K. Schum & Laut., Nachtr. Fl. Deutsch. Sudsee: 84 (1905). Type: New Guinea, *Schlechter* 14477 (holotype B).

Stems up to 35 cm tall, pinky to olive-brown. *Leaves* obliquely ovate, 3–5.5 cm long, 1–1.9 cm wide, green checkered with light green. *Inflorescence* up to 15 cm long. *Flowers* up to 4, with white sepals with maroon markings near apex and light green margins, white petals, a white lip and a pale orange spur; dorsal sepal obpyriform, 6.5 mm long; lateral sepals and petals oblique, lanceolate, 6 mm long; lip and spur 1 cm long; lip ovate, attenuate at obtuse apex, spur apex bilobed. (See fig. 2).

DISTRIBUTION: Ambae and Espiritu Santo. Also in New Guinea and the Solomon Islands.
HABITAT: Ridge-top forest, 600–900 m.
COLLECTIONS: *Cribb & Wheatley* 15 (K, PVNH) (sterile); *Wheatley* 31 (K) (sterile).

All collections of this species from Vanuatu, which we have examined, are sterile and therefore the description of the flowers is from Solomon Island collections.

2. V. salomonensis *Schltr.* in K. Schum. & Laut., Nachtr. Fl. Deutsch. Sudsee: 86 (1905). Type: Solomon Islands, *Guppy* 75 (holotype K!).
Vrydagzynea cheesemanii Ames in Journ. Arn. Arb. 14: 103 (1933); **synon. nov.**
  Type: Vanuatu, Malekula, *Cheesman* in RBG Kew 4 (holotype K!).

Stems up to 40 cm tall, green to brown. *Leaves* obliquely ovate, 6–8 cm long, 2–3 cm wide, midgreen. *Inflorescence* up to 12 cm long. *Flowers* c. 20, white with pale green-brown mottling on exterior of sepals; ovary and bracts pinkish; dorsal sepal oblong, c. 4 mm long; lateral sepals obliquely ovate, c. 4 mm long; petals obliquely ovate, 3 mm long; lip and spur c. 6 mm long; lip triangular, with recurved margins, spur conical-cylindrical, nectariferous. (See fig. 2).

DISTRIBUTION: Anatom, Banks Islands (Vanua Lava), Efate, Erromango, Espiritu Santo, Malekula, Pentecost and Tanna. Also in Bougainville and the Solomon Islands.
HABITAT: Bush and secondary forest, sea level to 600 m.
COLLECTIONS: *Bernardi* 13001 (G), 13179 (K, G, P), 13336 (K, P); *Braithwaite* in RSNH 2272 (P); *Cabalion* 574 (P), 1110 (PVNH), 2176 (P, PVNH) & 2246 (K, P, NOU); *Cheesman* in RBG Kew 4 (K); *Morat* 5476 (P); *Raynal* in RSNH 16178, 16183 & 16245 (P); *Wheatley* 251 & 354 (K, PVNH).

3. V. whitmeei *Schltr.* in Bull. Herb. Boiss. ser. 2, 6: 296 (1906). Type: Samoa, *Whitmee* s.n. (holotype B).

Stems up to 15 cm tall. *Leaves* lanceolate, acute, sometimes oblique, 2.7–5.5 cm long, 0.9–1.7 cm wide. *Inflorescence* up to 9 cm long. *Flowers* c. 10, whitish-yellow; dorsal sepal oblong, obtuse, c. 2.5 mm long; lateral sepals oblique-oblong, obtuse, c. 2.2 mm long; petals oblique-oblong, obtuse, c. 2 mm long; lip and spur c. 3 mm long; lip ovate-oblong, with reflexed margins, spur conical-cylindrical, minutely bilobed at apex.

DISTRIBUTION: Anatom, Banks Islands (Vanua Lava) and Erromango. Also in Samoa and Fiji.

HABITAT: Rain forest, 10–410 m.
COLLECTIONS: *Bernardi* 12972 (G); *Raynal* in RSNH 16244 (P); *Raynal &* *Schmid* in RSNH 16133 (P); *Wheatley* 358 (K, PVNH).

This species is similar to *Vrydagzynea albostriata* Schltr. from New Guinea and may prove to be conspecific.

14. **ZEUXINE** Lindley

Terrestrial or occasionally lithophytic. *Stems* leafy, arising from a creeping rhizome. *Leaves* ovate, petiolate. *Inflorescence* terminal with a few or many small flowers, pubescent. *Flowers* hardly opening; dorsal sepal and petals forming a hood; lateral sepals enclosing the base of lip; lip with a saccate base containing 2 glands, blade transversely widened, small, connected to the base by a claw; column short; pollinia 2.
A large genus of some 70 to 80 species from Africa to tropical and subtropical Asia and the Pacific Islands. Two species in Vanuatu.

Lip apex with a broad single-toothed sinus and auriculate lobes, forwards projecting …**2. Z. vieillardii**
Lip apex lacking sinus, reniform … … … … … … … … … … … … …**1. Z. erimae**

1. Z. erimae *Schltr.* in K. Schum. & Laut., Nachtr. Fl. Deutsch. Sudsee: 90 (1905). Type: New Guinea, *Schlechter* 13677 (holotype B).

Stem 5–15 cm tall, leafy. *Leaves* asymmetric, ovate, 4.5–10 cm long, 1.5–3 cm wide, green with a white central stripe; petioles c. 2 cm long, sheathing. *Peduncle* 14–30 cm tall; raceme 5–15 cm long, pubescent. *Flowers* 10–40, rather crowded, pubescent, with buff sepals and a white lip; pedicels c. 1 cm long, pubescent; sepals c. 4 mm long; lip c. 3 mm long, 3 mm wide, with a saccate base containing 2 glands, and a clawed, reniform to oblong apical blade.

DISTRIBUTION: Banks Islands (Vanua Lava), Erromango and Espiritu Santo. Also in New Guinea and Bougainville.
HABITAT: Bush and forest, on steep slopes, 80–800 m.
COLLECTIONS: *Cheesman* in RBG Kew 6 (K); *Cribb & Wheatley* 35 (K, PVNH); *Cuming* s.n. (G); *Raynal* in RSNH 161318 (K).

Zeuxine oblonga R.S. Rogers & C.T. White from Australia is similar and may prove to be conspecific.

2. Z. vieillardii (*Reichb. f.*) *Schltr.* in Engler, Bot. Jahrb. 39: 55 (1906).
Monochilus vieillardii Reichb. f. in Linnaea 41: 60 (1877). Type: New Caledonia, *Vieillard* 1311 (holotype P!).

Fig. 2. *Vrydagzynea salomonensis*. **A**, habit × ⅔; **B**, flower × 4; **C**, dorsal sepal × 6; **D**, lateral sepal × 6; **E**, petal × 6; **F**, anther cap × 8; **G**, lip × 6; **H**, cross section of lip × 6; **J**, column from below × 8. *V. argyrotaenia* **K**, flower × 2; **L**, dorsal sepal × 4; **M**, lateral sepal × 4; **N**, petal × 4; **O**, habit × ⅔; **P**, lip × 4; **Q**, lip cross section × 4; **R**, anther cap × 6; **S**, column from below × 6; **T**, pollinia × 6. **A** drawn from *Wickison* 119; **B–J** from *Wickison* 56 (Kew spirit no. 50883); **K–T** from *Wickison* 149 (Kew spirit no. 52089). All drawn by Sue Wickison.

Zeuxine daenikeri Kraenzl. in Viertelj. Nat. Ges. Zur. 74: 69 (1929). Type: New
Caledonia, *Daeniker* 163 (holotype Z).

Stem up to 14 cm tall, leafy. *Leaves* asymmetric, ovate, 3–5.5 cm long, 1–2 cm
long; petioles c. 1 cm long, sheathing. *Inflorescence* with peduncle up to 22 cm
long; raceme up to 20 cm long, pubescent. *Flowers* up to 25, rather crowded,
pubescent, cream to pale yellow-orange; sepals c. 4 mm long; lip c. 3 mm long, 2.5
mm wide, with a saccate base containing 2 glands, and a clawed, bilobed apical
blade with a broad, single-toothed sinus, the apical lobes auriculate and
projecting forwards.

DISTRIBUTION: Espiritu Santo and Tanna. Also in New Caledonia.
HABITAT: Bush close to rivers, sea level to 800 m.
COLLECTIONS: *Cheesman* 12 (K); *Raynal* in RSNH 16318 (K) & 16390 (P);
Raynal & Gillison in RSNH 16424 (K).

Raynal in RSNH 16318 (K) has much narrower leaves, 6.5–8 cm long, 0.8 cm
wide, than is typical for this species, however the flowers are identical with those
of *Zeuxine vieillardii*.

15. **SPIRANTHES** L.C. Richard

Terrestrial. Roots fleshy, fasciculate. *Stem* rather short, leafy. *Leaves* conduplicate,
fleshy. *Inflorescence* terminal, more or less densely spicate. *Flowers* small, arranged
spirally on the rhachis; sepals free; dorsal sepal erect, connivent with petals to
form a hood; lateral sepals oblique, more or less spreading; lip obscurely
trilobed; column rather short; pollinia 2.
A cosmopolitan genus of over 50 species. A single species in Vanuatu.

S. sinensis *(Pers.) Ames*, Orch. 2: 53 (1908).
Neottia sinensis Pers. in Syn. 2: 511 (1807). Type: China, collector not traced
(holotype UPS).
Spiranthes neocaledonica Schltr. in Engler, Bot. Jahrb. 39: 51 (1906). Type: New
Caledonia, *Schlechter* 15594 (holotype B).
For full synonymy see Garay and Sweet (1974).

Plant 8–40 cm tall. *Roots* fleshy, tuberous, cylindric, hairy. *Leaves* mostly basal,
varying greatly in size and shape, oblong-elliptic to linear–elliptic, up to 20 cm
long, 1 cm wide, commonly smaller. *Inflorescence* erect, slender; rhachis rather
densely or subdensely many-flowered. *Flowers* probably self-pollinating, white to
pale rose, rarely pink, semi-open; dorsal sepal ovate, up to 5 mm long; lateral
sepals oblong-lanceolate, c. 5 mm long; petals forming a hood with the dorsal
sepal, oblong-oblanceolate, slightly shorter than the dorsal sepal; lip obovate in
outline, obscurely trilobed, slightly longer than the sepals, lateral lobes semiovate, erect, midlobe suborbicular with an undulate-crispate margin, disc at base
on each side with a fleshy, subglobose callus.

DISTRIBUTION: Espiritu Santo. A quite remarkable distribution from
mainland Asia, S.E. Asia and the Malay archipelago to New Guinea, the Solomon
Islands, New Caledonia, Samoa and Australia.
HABITAT: Montane forest, 1000 m.
COLLECTION: No specimens seen, (Guillaumin, 1948).

Guillaumin cites a specimen, *I. Baker* 16, from Espiritu Santo, Mt. Tabwemasana, 3,000 ft as being *Spiranthes sinensis*.

16. **CRYPTOSTYLIS** R. Brown

Terrestrial. Leaves erect, arising from the rhizome, ovate, petiolate. *Inflorescence* arising separately from the rhizome; peduncle not sheathed by leaves, erect, long and slender. *Flowers* several, non-resupinate; sepals and petals narrow, spreading, with sepals longer than petals; lip erect, entire, widest near the base and tapering to the apex, strongly concave at the base, which surrounds the column; column very short; pollinia 4.

A genus of about 20 species from S.E. Asia to Australia and the Pacific Islands. A new genus record for Vanuatu, a single species being recorded.

C. arachnites (*Blume*) *Hassk.* in Cat. Bogor. 48 (1844).
Zosterostylis arachnites Blume in Bijdr. Fl. Ned. Ind.: 419 (1825). Type: Java, *Blume* s.n. (holotype L, isotype P).
Cryptostylis stenochila Schltr. in Engler, Bot. Jahrb. 39: 49 (1906). Type: New Caledonia, *Schlechter* 15596 (holotype B).

Plant 20–60 cm tall, with several hairy thick roots at the base. *Leaves* 1–4, ovate, acute, up to 17 cm long, 7.5 cm wide, pale green with a more or less distinct network of darker veins; petiole slender, 5–15 cm long, spotted with purple, not forming a sheath at the base. *Inflorescence* 10–30 cm long; peduncle covered with several overlapping sheaths. *Flowers* 8–12, with pale greenish or dull red-flushed sepals and petals; dorsal sepal 1.4–1.6 cm long, with edges inrolled; lateral sepals 1.4–1.6 cm long; petals 8–10 mm long; lip narrowed evenly to acute tip, 1.5–2 cm long, 5–7 mm wide at base, purplish red towards base, pale with deep red or purplish spots towards the apex with pubescent upper surface.

DISTRIBUTION: Espiritu Santo. Widely distributed from the Malay peninsula and Borneo to the Solomon Islands, New Caledonia, Fiji and Samoa.
HABITAT: In leaf litter in ridge-top montane forest, 900–1000 m.
COLLECTIONS: *Cribb & Wheatley* 50 (K, PVNH); *Veillon* 4021 (P).

17. **MEGASTYLIS** Schlechter

Terrestrial. Roots fleshy, fasciculate. *Leaves* grouped at the base of the stem. *Inflorescence* terminal, erect, racemose; peduncle sheathed in cataphylls; bracts often large. *Flowers* with dorsal sepal hooded; lip entire, often ovate; pollinia 2.

A small genus of about 7 species in New Caledonia and Vanuatu. A single species in Vanuatu.

M. gigas (*Reichb. f.*) *Schltr.* in Engler, Bot. Jahrb. 45: 379 (1911).
Caladenia gigas Reichb. f. in Linnaea 41: 56 (1877). Type: New Caledonia, *Vieillard* 1301 (holotype P!, isotype BM!, K!).
Lyperanthus gigas (Reichb. f.) Schltr. in Engler, Bot. Jahrb. 39: 44 (1906).
Lyperanthus sarasinianus Kraenzl. in Sarasin & Roux, Nov. Caled. 1: 78 (1914). Type: New Caledonia, *Sarasin* 352 (holotype B).

37

Plant 0.5–1.2 m tall. *Leaves* up to 8, distichous from base, narrowly lanceolate, up to 60 cm long, 2 cm wide, light green. *Inflorescence* racemose, densely many flowered; peduncle sheathed in cataphylls; bracts persistent, ovate 3–4 cm long. *Flowers* 6–30, white, lip may be edged with purple, scented; pedicel and ovary, 1–1.5 cm long, 6-ribbed; dorsal sepal large, hooded, ovate, 2.5–4 cm long, reflexed at apex; lateral sepals and petals lanceolate, 2.5–3 cm long; lateral sepals broader than petals, parallel with each other; lip ovate, c. 2.6 cm long, 1.4 cm wide, pubescent in middle, arched, reflexed at apex. (See plate 2c).

DISTRIBUTION: Anatom. Also in New Caledonia.
HABITAT: Rain forest, 100–200 m.
COLLECTIONS: *Milne* 245 (K); *Schmid* 3625 (P).

18. **CORYBAS** Salisbury

Terrestrial herbs arising from small tubers. *Leaf* solitary, ovate to orbiculate or cordate. *Flower* solitary on a short pedicel; dorsal sepal large and cucullate; lateral sepals and petals small and rudimentary or long and filiform; lip large, tubular in section in basal part, expanded and reflexed at apex, with 2 spurs; column small and erect; pollinia 4.

A genus of about 60 species from S.E. Asia to Australia and the Pacific islands. Three species in Vanuatu, *Corybas sp. nov.* being a new record.

1. Leaves 5 cm or more long; stems 11 cm or more long ... ... **3. C. sp. nov.**
   Leaves 4 cm long or less; stem 8 cm long or less   2
2. Leaves orbicular-cordate; lateral sepals and petals small and rudimentary
   ... ... **2. C. neocaledonicus**
   Leaves lanceolate-cordate; lateral sepals and petals long and filiform
   ... ... **1. C. mirabilis**

1. C. mirabilis (*Schltr.*) *Schltr.* in Fedde, Rep. Sp. Nov. 19: 22 (1923).
Corysanthes mirabilis Schltr. in Herb. Bull. Boiss. 6: 296 (1906). Type: Vanuatu, Anatom, *Morrison* s.n. (holotype B).

Stem 3–8 cm high, terete. *Leaf* lanceolate-cordate, apiculate-acuminate, 2.5–4.0 cm long, 1.5–2.5 cm wide, with slightly wavy margins, pale velvety green, tinged with purple on under surface. *Peduncle* 4–6 cm long. *Flower* with a lanceolate dorsal sepal, 2.0 cm long, 0.5 cm wide, maroon; lateral sepals and petals filiform, 2.5–3 cm long, pale green to white to pink; lip broadly oblong, obscurely trilobed, 1–1.3 cm long, 0.7 cm wide, maroon with white stripes, lateral lobes rounded, mid-lobe deflexed, acuminate, spurs conical; column c. 2.5 mm long.

DISTRIBUTION: Anatom. Also in the Solomon Islands.
HABITAT: Montane forest, c. 700 m.
COLLECTIONS: *Morrison* in RBG Kew 151 (K); *Raynal* in RSNH 16148 (P).

2. C. neocaledonicus (*Schltr.*) *Schltr.* in Fedde, Rep. Sp. Nov. 19: 23 (1924).
Corysanthes neocaledonica Schltr. in Engler, Bot. Jahrb. 39: 47 (1907). Types: New Caledonia, *Schlechter* 14918 (holotype B); New Caledonia, *McKee* 11472 (neotype CANB).

Stem 2–3 cm high. *Leaf* orbiculate-cordate, acute, 1.5–2.0 cm long, 1.5–2.0 cm wide, light green, with undulate margins. *Peduncle* c. 4 mm long. *Flower* red to dark red with a white mid-lobe to the lip; dorsal sepal obovate, 7–16 mm long, 5–9 mm wide; lateral sepals oblong-linear, falcate, 0.8–2 mm long; petals linear, 1.5–3 mm long; lip tubular in lower 4–6 mm, markedly curved downwards at tip and spreading into the 4–6 mm wide limb, the latter trilobed with lateral lobes rounded and curved outward and mid-lobe semicircular, 3–5 mm across, finely papillose on the inside mainly along the margin, spurs conical; column c. 2.8 mm long.

DISTRIBUTION: Anatom. Also in New Caledonia and Australia.
HABITAT: No information from Vanuatu. In New Caledonia it occurs in shade in damp forest, 400–600 m.
COLLECTION: *Morrison* in RBG Kew 57 (sterile) (K).

We have only seen a sterile specimen of this species and therefore the description of the flower is taken from van Royen (1983).

3. C. sp. nov.

Stem 11–16 cm long. *Leaf* ovate-cordate, acute, 4–5 cm long, 3.5–5 cm wide, glossy dark green above, shiny, paler green below.

DISTRIBUTION: Ambae and Espiritu Santo.
HABITAT: Submontane ridge forest and montane forest, 900–1290 m.
COLLECTIONS: *Cribb & Wheatley* 43 (K, PVNH) (sterile); *Wheatley* 53 (K) (sterile).

Unfortunately this species has not yet been collected in flower, but the leaf is quite distinct.

19. **MICROTIS** R. Brown

Terrestrial, arising from a small, more or less round or ovoid, subterranean tuber. *Leaf* solitary, arising directly from the tuber, more-or-less erect, hollow-terete. *Inflorescence* spicate, developed within the base of the hollow leaf and splitting and forcing its way out of the leaf. *Flowers* with concave dorsal sepal; lateral sepals narrower than the dorsal sepal, lanceolate to oblong, free; petals narrower, partly hidden by the dorsal sepal; lip oblong but sometimes dilated near the base, obtuse, truncate or emarginate, with entire or uneven margins, disc with 2 calli; column very short, with 2 lateral wings; pollinia 2.
A genus of about 10 species from S.E. Asia to Australia and the Pacific Islands. A single species in Vanuatu.

M. unifolia (*G. Forst.*) *Reichb. f.* in Beitr. Syst. Planz.: 62 (1871).
Ophrys unifolia G. Forst., Fl. Ins. Austr. Prodr.: 59 (1786). Type: New Zealand, *Forster* 167 (holotype BM!; isotype P!).
For full synonymy see Hallé (1977).

Plant 15–50 cm tall, slender. *Leaf* solitary, hollow-terete, up to 50 cm long, 3 mm wide. *Inflorescence* oblong-cyclindric, up to 25 cm long. *Flowers* dense, numerous, c. 40, pale or golden green; dorsal sepal ovate, cucullate, acute, c. 2 mm long; lateral sepals oblong, 1.6 mm long; petals linear-lanceolate, c. 1.2 mm long; lip oblong, truncate, c. 1.0 mm long, 0.8 mm wide, with 3 calli, one on either side near the base and one near the apex; margins erose.

DISTRIBUTION: Anatom. Widely distributed from the Malay archipelago to New Caledonia, Lord Howe Island and New Zealand.
HABITAT: Rainforest, 1000 m.
COLLECTION: *Cheesman* 12 (K).

Microtis parviflora R. Br. has been confused with *M. unifolia* (G. Forst.) Reichb. f. in the past but *M. parviflora* differs in having a tapering inflorescence and an oblong, rounded lip with entire margins, and only 2 calli.

20. **PERISTYLUS** Blume

Terrestrial. Stem arising from a tuber, erect. *Leaves* scattered along the stem or in a rosette; leaf bases sheathing. *Inflorescence* terminal, fairly long, racemose. *Flowers* small, many; dorsal sepal adhering to the petals and forming a helmet; lip trilobed, usually with a fleshy callus at the base of the lip, spurred; column short; stigma lobes sessile; pollinia 2.

A genus of 70 to 80 species from tropical and subtropical Asia to Australia and the Pacific Islands. Five species in Vanuatu, *Peristylus maculiferus*, *P. papuanus* and *P. wheatleyi* being new records.

1. Spur shorter than sepals, 2–2.5 mm long, globular   **2. P. novoebudarum**
   Spur longer than sepals, 4.5–9 mm long, fusiform 2
2. Leaves in a terminal rosette; base of lip with an erect horn
   ... ... **4. P. stenodontus**
   Leaves scattered in apical half of stem; base of lip without an erect horn
   3
3. Leaves 12–30 cm long, 3.7–6 cm wide; base of lip with a fleshy trilobed
   callus   ... ...**5. P. wheatleyi**
   Leaves 9–15.5 cm long, 1.3–2.4 cm wide; base of lip with a transverse or
   rounded callus 4
4. Lateral lobes of lip c. 2 mm long; base of lip with a transverse callus
   ... ...**1. P. maculiferus**
   Lateral lobes of lip c. 4 mm long; base of lip with a rounded callus
   ... ...**2. P. papuanus**

1. P. maculiferus (*C. Schweinf.*) *Renz & Vodonaivalu* **comb. nov.**
Habenaria maculifera C. Schweinf. in Bull. Bishop Mus., Honolulu 141: 18 (1936).
  Type: Fiji, *A.C. Smith* 1911 (holotype K!).

Plant up to 50 cm tall. *Leaves* scattered towards apex of stem, lanceolate, acute, 9, 10–15.5 cm long, 1.3–2.4 cm wide; leaf sheaths may be spotted. *Inflorescence* 25 cm long; raceme 15 cm long; bracts 8 mm long. *Flowers* c. 20, green with center violet; sepals ovate, 2.5 mm long; petals obliquely rhombic, 1.6 mm long; lip trilobed, with midlobe shorter than lateral lobes, midlobe 1 mm long, lateral lobes 2 mm long, spur fusiform, 4 mm long.

DISTRIBUTION: Banks Islands (Vanua Lava). Also in Fiji.
HABITAT: Rainforest, 300 m.

COLLECTION: *Veillon* 5514 (NOU).

2. P. novoebudarum *F. Muell.*, Phyt. N. Hebrid.: 22 (1873). Type: Vanuatu, Anatom, *Campbell* s.n. (holotype MEL).
Habenaria physoplectrus Reichb. f. in Linnaea: 17 (1877); **synon. nov.** Type: Vanuatu, Ánatom, *MacGillivray* 27 in part (holotype W, isotypes G!, P!).
Habenaria ngoyensis Schltr. in Engler, Bot. Jahrb. 3: 34 (1906); **synon. nov.** Type: New Caledonia, *Schlechter* 15283 (holotype B).
Peristylus ngoyensis (Schltr.) N. Hallé, Fl. Nouv. Caled. 8: 550 (1977); **synon. nov.**
Peristylus physoplectrus (Reichb. f.) N. Hallé, Fl. Nouv. Caled. 8: 550 (1977); **synon. nov.**

Plant up to 50 cm tall. *Tuber* narrow, ellipsoid–cylindrical. *Leaves* in terminal rosette, lanceolate, acute, c. 5, 11–17 cm long, 2–5 cm wide, lowest leaves c. 3 cm long, 0.5–1 cm wide; petioles tubular, sheathing. *Inflorescence* c. 30 cm long; raceme c. 20 cm long; bracts c. 1 cm long. *Flowers* 15–50, greenish yellow to brownish-yellow; sepals and petals ovate, c. 3 mm long; lip trilobed, midlobe slightly shorter than lateral lobes, lateral lobes c. 3 mm long, base of lip with an erect horn c. 0.5 mm long, spur globular, 2–2.5 mm long.

DISTRIBUTION: Anatom, Banks Islands (Vanua Lava), Efate, Erromango, Espiritu Santo, Pentecost and Tanná. Also in New Caledonia.
HABITAT: Rain forest, 160–1000 m.
COLLECTIONS: *Bernardi* 13172, 13334 (G, P) & 13359 (P); *Cabalion* 1359, 1412 & 1932 (NOU, P); *Cheesman* s.n. (K); *Kajewski* 169 & 942 (K); *MacGillivray* 27 (G, P); *Morat* 5210, 5927 (P) & 5907 (NOU, P); *Morrison* in RBG Kew 60, 61, 62, 127, 128 & 129 (K); *Veillon* 4020 (P).

3. P. papuanus *(Kraenzl.) J.J. Smith* in Nova Guinea 12: 3 (1913).
Habenaria papuana Kraenzl. in Engler, Bot. Jahrb. 18: 188 (1894). Type: New Guinea, *F. Hellweg* 585 (holotype B).
Habenaria cyrtostigma Schltr. in Fedde, Rep. Sp. Nov. 9: 83 (1910). Types: Samoa, *Rechinger* 732 (syntype W; isosyntype G); *Rechinger* 1802 (syntype W); *Rechinger* 1146 (syntype W).

Plant up to 60 cm tall. *Leaves* scattered towards apex of stem, lanceolate, acute, 7, 9–12 cm long, 1.9–2.0 cm wide. *Inflorescence* 28 cm long; raceme 16 cm long; bracts 7 mm long. *Flowers* green, c. 20; ovary and pedicel 8–10 mm long; dorsal sepal ovate, 2.6 mm long; lateral sepals obliquely ovate, 3 mm long; petals obliquely rhombic, 2.2 mm long; lip trilobed, with midlobe shorter than lateral lobes, midlobe 1.2 mm long, lateral lobes 4 mm long, base of lip with a rounded callus, 0.8 mm long, spur fusiform, 5.5 mm long. (See fig. 3).

DISTRIBUTION: Anatom. Also in New Guinea, the Solomon Islands, Fiji and Samoa.
HABITAT: Rain forest.
COLLECTION: *Morrison* in RBG Kew 129 (K).

4. P. stenodontus *(Reichb. f.) Renz & Vodonaivalu* **comb. nov.**
Habenaria stenodonta Reichb. f. in Linnaea 42: 17 (1876). Type: Vanuatu, Anatom, *MacGillivray* 27 in part (holotype W).

42

Leaves in a rosette held well above the ground, lanceolate, acute, 12 cm long, 2.5 cm wide. *Inflorescence* c. 40 cm long; raceme c. 25 cm long; bracts 6–11 mm long. *Flowers* 20–30; sepals ovate, c. 3.5 mm long; petals obliquely rhombic, c. 3.5 mm long; lip trilobed, with midlobe shorter than lateral lobes, midlobe c. 2.2 mm long, lateral lobes c. 3 mm long, base of lip with a globular callus, c. 1 mm long and an erect horn c. 1 mm long, spur fusiform, 4.5–9 mm long.

DISTRIBUTION: Anatom and Banks Islands (Vanua Lava).
HABITAT: Bush, 30 m.
COLLECTIONS: *Cabalion* 1932 (NOU); *Cheesman* in RBG Kew 2 (K); *Green* in RSNH 1160 (K); *Morat* 7490 (P); *Wheatley* 351 (K, PVNH).

MacGillivray 27 is a mixed collection, containing the types of *Peristylus stenodontus* and *P. physoplectrus* (Renz, pers. comm.).

5. P. wheatleyi *Cribb & B. Lewis* in Orchid Rev. in press (1989). Type: Vanuatu, Ambae, *Wheatley* 33 (holotype K!).

Stem 120 cm tall, stout, light green, leafy in the upper half and covered in tubular sheaths in the lower half. *Leaves* c. 6, lanceolate to oblanceolate, acuminate, spreading, 12–30 cm long, 3.7–6 cm wide, thin-textured, dark green with darker venation, shortly petiolate, with a tubular-funnel-shaped sheathing base. *Inflorescence* a many-flowered cylindrical raceme, up to 55 cm long; raceme 29 cm long; bracts linear-lanceolate, acuminate, 1–1.8 cm long. *Flowers* c. 60, bright green with 3 purple spots on the lip and with white petals mottled with purple; pedicels c. 1.3–1.5 cm long; dorsal sepal concave, ovate, acute, 3.5 mm long, 2 mm wide; lateral sepals reflexed, obliquely oblong-ovate, apiculate, 5 mm long, 2 mm wide, slightly keeled on the outer surface; petals adnate to the dorsal sepal to form a hood over the column, 4.5 mm long, 2.5 mm wide; lip trilobed, 5 mm long, 10 mm wide, with midlobe shorter than lateral lobes, deflexed, with an obscure trilobed fleshy basal callus at the mouth of the spur, lateral lobes spreading, upcurved at apex, lanceolate, acute, falcate, 5–7 mm long, midlobe almost triangular, 2.5 mm long, spur slender, cylindrical, parallel to the ovary, upcurved or deflexed, 8–10 mm long; column very short, 0.8 mm long. (See fig. 4).

DISTRIBUTION: Anatom, Ambae and Banks Islands (Vanua Lava).
HABITAT: Montane *Weinmannia-Myrtaceae* forest, 750–1000 m.
COLLECTIONS: *Green* in RSNH 1160 (K, P); *Wheatley* 33 & 359 (K, PVNH).

The flowers of *Peristylus wheatleyi* are similar in shape to those of *P. triaena* (Schltr.) P.F. Hunt from New Guinea, but that is a much smaller plant reaching only 35 cm in height and with much smaller leaves and flowers. *P. remotifolius* J.J. Smith also from New Guinea has similar flowers as well but is a much smaller plant with shorter narrower lanceolate leaves, fewer smaller flowers, a shorter spur and stalked upcurved stigma lobes.

Fig. 3. *Peristylus papuanus*. **A**, habit × ⅔; **B**, column × 12; **C**, flower × 6; **D**, lip × 6; **E**, dorsal sepal × 8; **F**, lateral sepal × 8; **G**, petal × 8. **A** drawn from *Hunt* 2279; **B–G** from *Dennis* 2279 (Kew spirit no. 28866) by Sue Wickison.

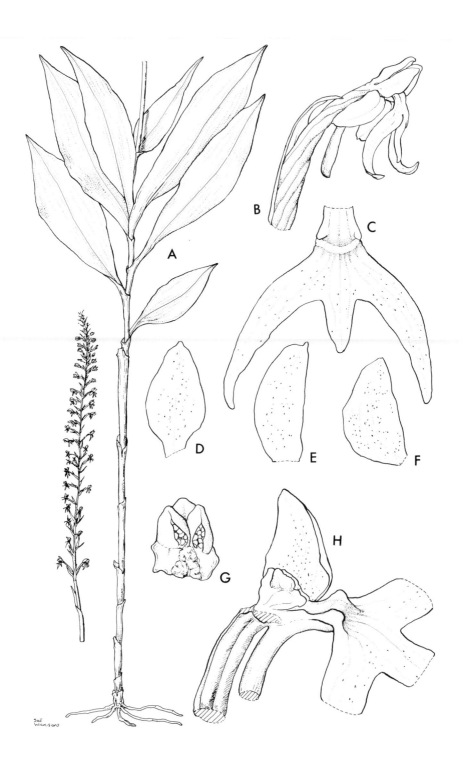

44

21. **HABENARIA** Willdenow

Terrestrial. Stem arising from a tuber, erect. *Leaves* scattered along stem or in a rosette; leaf bases sheathing. *Inflorescence* terminal, long, racemose. *Flowers* small to large, few to many; sepals free; petals similar to sepals or bilobed, often adnate to the dorsal sepal to form a hood over the column; lip trilobed, spurred; column short; stigma lobes are distinctly stalked; pollinia 2.

A large genus of 600–700 species in the tropics and warm temperate regions of the world. A single species in Vanuatu, *Habenaria novaehiberniae* is a new record.

H. novaehiberniae *Schltr.* in K. Schum. & Laut., Nachtr. Fl. Deutsch. Sudsee.: 79 (1905). Type: New Ireland, *Schlechter* 14698 (holotype B).

Stem up to 60 cm high, green tinged with dull purple. *Leaves* in an apical rosette, 6–7, ovate-lanceolate, 11–14 cm long, 3.5–3.8 cm wide, with bases sheathing stem; leaf sheaths may be spotted. *Inflorescence* up to 40 cm long; bracts lanceolate, 1.5–2 cm long. *Flowers* 10–18, whitish green to green; pedicel and ovary 2.5 cm long; dorsal sepal ovate, 9 mm long; lateral sepals obliquely ovate, 9 mm long; petals bipartite with linear lobes, posterior lobe c. 8 mm long, slightly broader than the anterior lobe, anterior lobe c. 10.5 mm long; lip trilobed with linear lobes, lateral lobes longer than the midlobe, c. 18 mm long, curled, midlobe c. 7 mm long, spur filiform, shorter than the ovary, up to 20 mm long.

DISTRIBUTION: Efate, Erromango and Tanna. Also in New Guinea and the Solomon Islands.
HABITAT: Dense forest, 200–750 m.
COLLECTIONS: *Cabalion* 1628 (NOU, P, PVNH) & 1725 (NOU, PVNH); *Green* in RSNH 1077 (K, P); *Hallé in RSNH 6391 (K, P); Morrison* in RBG Kew 62 (K).

22. **NERVILIA** Commelin ex Gaudichaud

Terrestrial, arising from a more or less round tuber, which has short roots. *Inflorescence* erect, produced before the leaf appears, elongating after fertilization of the ovules, terminating in a 1–few-flowered raceme. After the flowers mature a solitary leaf arises, on a separate stalk. *Leaf* plicate, cordate, ovate or orbicular-reniform. *Flower* with sepals and petals similar and free, long and narrow; lip entire or trilobed; column rather long, dilated at apex; pollinia 2.

The whole flowering cycle, from the first emergence of the flowering stem to the dispersal of the seed is brief, occupying only a few weeks, after which the stem collapses and quickly rots away. The leaf, however, persists for several months.

A genus of about 80 species in the tropics and subtropics of the Old World. Two species in Vanuatu.

Fig. 4. *Peristylus wheatleyi.* **A**, habit × ¼; **B**, flower × 3; **C**, lip × 6; **D**, dorsal sepal × 6; **E**, lateral sepal × 6; **F**, petal × 6; **G**, column × 14; **H**, side view lip and column × 6. Drawn from *Wheatley* 33 (Kew spirit no. 53350) by Sue Wickison.

Leaf cordate, c. 8 cm long, 8 cm wide; inflorescence 3- or more-flowered; lip trilobed ... ... **1. N. aragoana**
Leaf orbicular-reniform, sinuately 7–9-lobed, c. 3.5 cm long, 5 cm wide; flowers solitary; lip not lobed ... ... **2. N. crociformis**

1. N. aragoana *Gaud.* in Freyc., Voy. Bot.: 422, t.35 (1826). Type: Mariana Islands, *Gaudichaud* s.n. (holotype P!).
For full synonymy see Seidenfaden (1978).

Leaf erect, cordate, acuminate or acute, 8 cm long, 8 cm wide; petiole 15–20 cm long. *Inflorescence* 15–30 cm tall, sheathed by bracts. *Flowers* 3–7, with pale green sepals and petals and a white lip, probably self-pollinating; sepals and petals linear, narrowly ovate, acuminate, 2–3.5 cm long; lip trilobed, 10.5 mm long, 6.5 mm wide, lateral lobes erect, short and broad, obliquely triangular, midlobe broad-ovate.

DISTRIBUTION: Erromango, Espiritu Santo and Tanna. Widely distributed from Asia and the Malay archipelago to the Mariana Islands, New Guinea, the Solomon Islands, New Caledonia, the Horn Islands, Fiji, Samoa, Niue Island and Australia.
HABITAT: Secondary forest, on limestone, 100–190 m.
COLLECTIONS: *Cribb & Wheatley* 4 (K, PVNH); *Hoock* s.n. (P); *Morat* 5919 (P); *Raynal* in RSNH 16294 (K, P).

2. N. crociformis (*Zoll. & Mor.*) *Seidenf.* in Dansk Bot. Arkiv 4: 151 (1978).
Bolborchis crociformis Zoll. & Mor. in Moritzi, Syst. Verz. Pl. Zoll.: 89 (1846). Type: Java, *Zollinger* 762 (holotype P).
Pogonia crispata Blume in Mus. Bot. Lugd. Bat. 1: 32 (1849). Type: Java, *Blume* s.n. (holotype L).
Nervilia crispata (Blume) Schltr. ex Kraenzl. in K. Schum. & Laut., Fl. Deutsch. Schutzgeb.: 240 (1901).
Nervilia fimbriata Schltr. in K. Schum. & Laut., Nachtr. Fl. Deutsch. Sudsee: 82 (1905). Type: New Guinea, *Schlechter* 13795 (holotype B).
For full synonymy see Seidenfaden (1978).

Leaf erect, orbicular-reniform, sinuately 7–9-lobed, up to 3.5 cm long, 5 mm wide, slightly pubescent; petiole up to 3 cm long. *Inflorescence* up to 13 cm tall, sheathed by bracts. *Flower* solitary; sepals linear, narrowly ovate, acuminate, c. 1.5 cm long; lip not lobed, c. 1 cm long, 0.6 cm wide, disc hairy in middle, with apical margin fimbriate.

DISTRIBUTION: Anatom. Widely distributed from the Malay archipelago to New Guinea, New Caledonia, Samoa and Australia.
HABITAT: Lowland forest.
COLLECTIONS: *Morrison* in RBG Kew 58 & 59 (K) (sterile).

Both collections from Vanuatu which we have seen are sterile and therefore the description of the inflorescence is from New Guinea collections.

23. **DIDYMOPLEXIS** Griffith

Terrestrial. Saprophytic. Rhizome large and fleshy. *Inflorescence* a terminal raceme on an erect stem; pedicels elongating rapidly after fruit set. *Flowers* with sepals and petals joined in a short tube at the base but with petals joined to dorsal sepal for a greater distance than to the lateral ones; lip joined to column-foot, free from sepals and petals; column long, slightly widened at apex, foot short; pollinia 4.

A genus of about 20 species in the Old World Tropics. A single species in Vanuatu.

D. micradenia (*Reichb. f.*) *Hemsley*, in Journ. Linn. Soc. 20: 311 (1883).
Epiphanes micradenia Reichb. f. in Seem. Fl. Vit.: 295 (1868): Type: Fiji, *Seemann* 610 (holotype K!).
Didymoplexis minor J.J. Smith in Bull. Inst. Bogor. 7: 1 (1900); **synon. nov**. Type: Java, *J.J. Smith* 74 (holotype L).
Didymoplexis neocaledonica Schltr. in Engler, Bot. Jahrb. 39: 50 (1906). Type: New Caledonia, *Schlechter* 15748 (holotype B).
Didymoplexis fimbriata Schltr. in Engler, Bot. Jahrb. 56: 449 (1921); **synon. nov**. Type: Palau Islands, *Ledermann* 14572 (holotype B).

Rhizome to 8 cm long. *Stems* erect, 5–12 cm long. *Inflorescence* terminal, racemose. *Flowers* few, pale brown, olive or pinkish, with a yellowish white lip, spotted with pink; sepals c. 6.5 mm long; lip erect, with margins rolled inwards enclosing the column, obscurely trilobed, widest towards apex, 5.5 cm long, 3.2 mm wide, with central disc of midlobe raised, fleshy and with 3 verrucose calli. (See fig. 5).

DISTRIBUTION: Espiritu Santo and Malekula. Widely distributed from Java and the Palau Islands to New Caledonia, the Horn Islands, Fiji and Samoa.
HABITAT: Rain forest and secondary forest on raised coral, c. 100 m.
COLLECTIONS: *Cribb & Wheatley* 3 (K); *Hallé* in RSNH 6353 & 6354 (P).

This species is similar to *Didymoplexis pallens* Griff. from India but it differs in the shape of the lip.

24. **GASTRODIA** R. Brown

Terrestrial. Saprophytic. Rhizome tuberous, more or less horizontal. *Stems* tall or short, erect, often elongating after fertilization, leafless. *Inflorescence* terminal, racemose, of 1 to many flowers. *Flowers* with sepals and petals joined to form a 5-lobed tube which may or may not be gibbous at the base, and which may be split between the lateral sepals; petals smaller than sepals; lip shorter than sepals, entire or trilobate, disc with 2 keels and sometimes glandular calli at base; column fairly long with a distinct, but often quite short foot; pollinia 2.

A genus of about 20 species from E. and S.E. Asia to Australia, New Zealand and the Pacific Islands. A single species in Vanuatu.

G. cunninghamii *Hook. f.*, Fl. Nov. Zeal. 1: 251 (1855). Type: New Zealand, *Colenso* s.n. (holotype K!).

Gastrodia orobanchoides F. Muell., Phyt. N. Hebrid.: 22 (1873); **synon. nov.** Type: Vanuatu, Erromango, *Fraser* s.n. (holotype MEL).
Gastrodia leucopetala Colenso in Trans. N.Z. Inst. 18: 268 (1886). Type: New Zealand, *Colenso* s.n. (holotype K!).

Inflorescence up to 60 cm long, many-flowered, raceme up to 30 cm long, slightly hairy; bracts lanceolate, acuminate. *Flowers* c. 20; sepals white, grey inside; petals white; lip white with a yellow tip; with a strong odour; pedicel and ovary short, c. 5 mm long; sepals c. 1 cm long, fused into a tube, gibbous; lip oblong, with undulating margins, with a central ridged callus which extends to the reflexed tip, with the base joined to the perianth tube and having thick orange, lateral calli; column pale yellow, very short and curled over the stigma; as the flowers open the column is forced forwards and downwards until the pollinia touch the stigma, thus ensuring self-pollination.

DISTRIBUTION: Erromango. Also in New Zealand, Stewart Island and Chatham Island.
HABITAT: No information from Vanuatu. In New Zealand it is often associated with *Nothofagus* spp. It usually occurs in shaded areas but also occurs in the middle of cleared tracks and on banks.
COLLECTION: *Fraser* s.n. (MEL) (not seen).

The description is from New Zealand specimens and Mueller's description of *Gastrodia orobanchoides*; more information is needed on this species in Vanuatu.

25. **EPIPOGIUM** J.G. Gmelin

Terrestrial. Saprophytic. Rhizome fleshy. *Inflorescence* ephemeral, terminal on an erect fleshy stem. *Flowers* non-resupinate, with sepals and petals of a more or less equal length, narrow; lip concave with minutely warty ridges, spurred; column short; pollinia 2.
A small genus of 2 species from Europe and Asia to New Caledonia and Australia. A single species in Vanuatu.

E. roseum (*D. Don*) *Lindley* in Journ. Linn. Soc. 1: 177 (1878).
Limodorum roseum D. Don, Prodr. Fl. Nep.: 30 (1825). Type: Nepal, *Wallich* (holotype BM; isotype K!).
For full synonymy see Hallé(1977).

Plant 5–60 cm tall, leafless, brittle and pale buff to dull yellow in all its parts. *Rhizome* embedded only shallowly in humus, a horizontal ovoid tuber up to 6 cm long, 4 cm wide, but usually much less, lacking roots. *Peduncle* hollow, sheathed by bracts; rhachis usually nodding at first, 2–20 cm long. *Flowers* few to many, not spreading, cream to yellow, self-fertilising; pedicel 3–5 mm long; ovary swollen, 5–8 cm long; sepals narrowly oblong, 7–12 mm long; petals shorter and broader

---

Fig. 5. *Epipogium roseum.* **A**, habit × ⅔; **B**, flower × 3; **C**, column and spur × 6; **D**, anther cap × 8; **E**, lip × 4; **F**, dorsal sepal × 4; **G**, petal × 4; **H**, lateral sepal × 4. *Dipodium punctatum* var. *squamatum.* **J**, habit × ⅔; **K** inflorescence; **L**, lip × 4. *Didymoplexis micradenia.* **M**, habit × ⅔; **N**, flower × 3; **O**, lip × 6. A–H drawn from *Bregulla* s.n.; **J** from *Hallé* in RSNH 1610; **K** from slide; **L** from *Hallé* in RSNH 16157; **M** & **O** from *Cribb & Wheatley* 3; **N** from slide. All drawn by Sue Wickison.

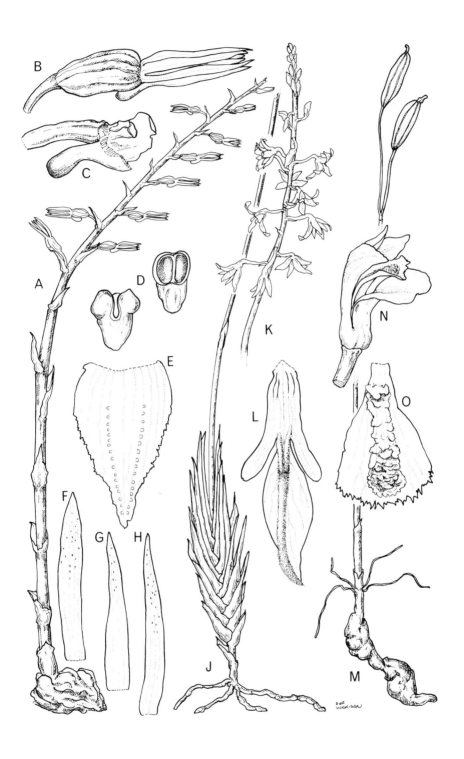

than sepals; lip concave to more or less enclose the column, about the same length as the sepals, triangular-ovate, upper surface minutely warty in 2 rows, decurved at apex, spur cylindric, obtuse, 2–4 mm long, pointed backwards; column 2–2.5 mm long. (See fig. 5).

DISTRIBUTION: Erromango and Malekula. Widely distributed from Africa, Asia and the Malay archipelago to New Guinea, New Caledonia and Australia.

HABITAT: Variety of habitats from bush to rain forest and from the coast to hills, sea level to 190 m.

COLLECTIONS: *Cheesman* in RBG Kew 5 (K); *Hallé* in RSNH 6355 (K, P, PVNH); *Raynal* in RSNH 16299 (P); *Sam* 287 (K, P, PVNH).

Epipogium roseum has an interesting life-history (Docters van Leeuwen, 1937), the growth of the inflorescence being very rapid, its whole life, even including the development of fruit and seeds, occupying only a few days. There is no rostellum, and the pollen comes into contact with the stigma one or two days before the flowers open. On the third of fourth day after the flowers open, the fruit dehisces and the seeds are scattered. The embryo is unusually small (consisting of 8 cells only) and the seeds are among the lightest of all orchid seeds.

26. **ACANTHEPHIPPIUM** Blume

Terrestrial herbs. *Pseudobulbs* few noded, fleshy, subcylindric, green, covered by sheaths when immature, 2-leaved at apex. *Leaves* suberect, large, plicate. *Inflorescence* lateral, shorter than the leaves, few-flowered. *Flowers* large, urn-shaped; sepals fleshy, connate with free apices; lateral sepals forming a chin-like mentum with the column-foot; petals included within the sepals, much narrower; lip trilobed, articulate with the column-foot, long-clawed; column with a long foot; pollinia 8.

A genus of about 15 species from tropical Asia to the Pacific Islands. A new genus record for Vanuatu, a single species being recorded.

A. papuanum *Schltr.* in Fedde, Rep. Sp. Nov., Beih. 1: 371 (1912). Type: New Guinea, *Schlechter* 16894 (holotype B).

Stems including the pseudobulb up to 26.5 cm long. *Leaves* plicate, lanceolate, up to 65 cm long, 11 cm wide, longly petiolate. *Inflorescence* up to 25 cm long, erect, 3–6-flowered; bracts up to 3.5 cm long. *Flowers* 3.5–4.5 cm long, yellow, lined and flushed with red; mentum 1.3–1.4 cm long; lip trilobed, lateral lobes obliquely subquadrate, mid-lobe obovate, callus obsure, of 3 interrupted ridges.

DISTRIBUTION: Banks Islands (Vanua Lava). Also in New Guinea and the Solomon Islands.

HABITAT: Terrestrial in deep leaf litter in primary forest; up to 400 m.

COLLECTION: *Wheatley* 350 (K, PVNH).

Acanthephippium vitiense L.O. Williams from Fiji is similar and may be conspecific.

27. **CALANTHE** R. Brown

Terrestrial herbs with thick hairy roots. *Stems* pseudobulbous, often clustered. *Leaves* plicate, petiolate. *Inflorescence* lateral, erect, unbranched, few–many-flowered. *Flowers* often showy, usually turning blue when damaged; sepals and petals similar, free; lip 3- or 4-lobed, adnate to column at base, spurred; column short, fleshy; pollinia 8.

A genus of about 100 species in the tropics and warm temperate regions distributed from Africa to Asia and the Pacific Islands, with a single species in the tropical Americas. Three species in Vanuatu.

1. Lip deeply four-lobed ... ... **3. C. triplicata**
   Lip three-lobed 2
2. Flowers yellow or orange-yellow; spur 5–7 mm long, appressed to ovary; lip with bilobed basal callus ... ... **2. C. langei**
   Flowers green to white; spur c. 1 cm long, not appressed to ovary; lip without a callus ... ... **1. C. hololeuca**

1. C. hololeuca *Reichb. f.* in Seem., Fl. Vit.: 298 (1868). Type: Fiji, *Seemann* 607 (holotype W; isotype K!).
Calanthe neohibernica Schltr. in K. Schum. & Laut., Nachtr. Fl. Deutsch. Sudsee: 142 (1905). Type: New Ireland, *Schlechter* 14707 (holotype B).
Calanthe vaupeliana Kraenzl. in Notizbl. Bot. Gart. Berlin 5: 111 (1909). Type: Samoa, *Vaupel* 358 (isotype K!).
Calanthe neocaledonia Rendle in Journ. Linn. Soc. Bot. 45: 251 (1921). Types: New Caledonia, *Compton* 1409 & 1609 (syntypes BM!).

Leaves narrowly lanceolate, up to 65 cm long, 7.5 cm wide. *Inflorescence* c. 35 cm long; rhachis c. 12 cm long; bracts deciduous. *Flowers* white, sometimes with green side-lobes to lip; sepals and petals 10–12 mm long; lip trilobed, 7 mm long, lateral lobes auriculate, spur slightly S-shaped, c. 1 cm long.

DISTRIBUTION: Anatom, Banks Islands (Vanua Lava) and Tanna. Also in New Ireland, the Solomon Islands, the Horn Islands, Fiji and Samoa.
HABITAT: Primary rain forest, 260–500 m.
COLLECTIONS: *Bernardi* 13133 (P); *Morrison* in RBG Kew 119, 120, 122 & 123 (K); *Veillon* 5534 (P).

2. C. langei *F. Muell.* in Wing's South Sci. Rec.: 1 (1885). Type: New Caledonia, cult. *Lange* s.n. (holotype MEL).
Calanthe ventrilabium Reichb. f. in Seem., Fl. Vit.: 298 (1898); **synon. nov.** Type: Fiji, *Seemann* 606 (holotype K!).
Calanthe chrysantha Schltr. in K. Schum. & Laut., Nachtr. Fl. Deutsch. Sudsee: 141 (1905). Type: New Guinea, *Schlechter* 14494 (holotype B).
Calanthe angustifolia auct. non (Blume) Lindley 1830.
Calanthe lyroglossa auct. non Reichb. f. 1878.

Leaves 40–100 cm long, 5–9.5 cm wide. *Inflorescence* c. 35 cm long, densely many-flowered; rhachis 6–7 cm long; bracts deciduous. *Flowers* yellow to orange; sepals and petals c. 1.1 cm long; lip trilobed, 5.5–6.5 mm long, lateral lobes auriculate, callus of 2 short low fleshy ridges, spur clavate, parallel to ovary, 5–7 mm long. (See plate 2d).

DISTRIBUTION: Anatom, Efate, Erromango, Espiritu Santo and Tanna. Also in New Guinea, Bougainville, the Solomon Islands, Fiji and Samoa.

HABITAT: Rain forest, 150–1000 m.

COLLECTIONS: *Cabalion* 716 & 2121 (P); *Cheesman* 10 (K); *Green* in RSNH 1174, 1226 (K, PVNH) & 1293 (K); *Hallé* 68 (PVNH); *Raynal* in RSNH 16156 (K, P); *Chew Wee-Lek* in RSNH 68 (K, P).

3. C. triplicata (*Willemet*) *Ames* in Phillipp. Journ. Sc., Bot. 2: 326 (1907).

Flos triplicata Rumph., Herb. Amb. 6: 115. t.52 f.2. nom. illeg. pre 1753.

Orchis triplicata Willemet in Usteri, Ann. Bot. 18: 52 (1796). Type: Amboina, *Rumphius* s.n. (holotype L).

Limodorum veratrifolium Willd., Sp. Pl. ed. 4, pars 1: 122 (1805). Type: as for *O. triplicata*

Calanthe veratrifolia (Willd.) R. Br. in Bot. Reg. 9: t. 720 (1823).

Calanthe furcata Batem. ex Lindley. in Bot. Reg. 24: misc. p. 28 (1838). Type: Philippines, *Cuming* s.n. (holotype K!).

Calanthe angraeciflora Reichb. f. in Linnaea 41: 75 (1877). Type: New Caledonia, *Deplanche* 114 (holotype P!).

Calanthe veratrifolia var. *cleistogama* Schltr. in Fedde, Rep. Sp. Nov., Beih. 1: 380 (1912). Type: New Guinea, *Schlechter* 19054 (holotype B).

Calanthe triplicata var. *angraeciflora* (Reichb. f.) N. Hallé, Fl. Nouv. Caled. 8: 230 (1977). Type: as for *Calanthe angraeciflora*.

Calanthe quaifei Rolfe, nom. ined. in Herb. K.

For full synonymy see Garay & Sweet (1974).

Leaves up to 75 cm long, 10 cm wide, dark green with paler venation. *Inflorescence* to 125 cm long, many-flowered; bracts persistent, lanceolate. *Flowers* rarely cleistogamous, white with a yellow callus, pubescent on outer surfaces; sepals and petals 1.4 cm long or more; lip four-lobed, c. 2 cm long, lobes subequal, callus of 3 tapering verrucose ridges, spur decurved, cylindrical, more than 2 cm long. (See plate 2b).

DISTRIBUTION: Anatom, Efate, Espiritu Santo, Malekula, Pentecost and Tanna. Widely distributed from Madagascar, Asia and S.E. Asia to the Malay archipelago, New Guinea, the Solomon Islands, Lord Howe Island and Australia.

VERNACULAR NAME: Natapak.

HABITAT: Dense rain forest to open grasslands and bush, sea level to 920 m.

COLLECTIONS: *Bernardi* 13128 (K, P); *Cabalion* 1358 (PVNH); *Cheesman* in RBG Kew 12, 14 (K) & A10 (BM); *Cribb & A. Morrison* 1810 (K); *Cribb & Wheatley* 42 & 70 (K, PVNH); *Hallé* in RSNH 6367 & 6388 (P); *McKee* 32697 (P); *Morat* 5187 (P); *Morrison* in RBG Kew 51 & 54 (K); *Quaife* 7 (K); *de la Rüe* s.n. (P); *Chew Wee-Lek* in RSNH 237 (K); *Wheatley* 130 (K, PVNH).

Hallé (1977) recognises the New Caledonian plants as a distinct variety, var. *angraeciflora*. However, we consider that *Calanthe triplicata* is so variable throughout its wide distribution that this variety cannot be recognized as distinct.

Wheatley (pers. comm.) has noted that the flowers close up at night.

28. **PHAIUS** Loureiro

Terrestrial. Leaves 4–7, narrowly obovate, acuminate, medium to large in size, plicate, petiolate. *Inflorescence* basal or lateral, erect, loosely racemose. *Flowers* rather large and showy, turning blue when damaged; sepals and petals free; lip free, obscurely trilobed, with lateral lobes inrolled to form a tube which encompasses the column, shortly spurred at the base; column long; pollinia 8.

A genus of about 40 species in the Old World Tropics. Three species in Vanuatu, *Phaius robertsii* being a new record.

1. Stem thickened into a short, many-jointed pseudobulb; peduncle arising from the base of the pseudobulb; flowers pale rose to brown-orange, with a white lip with tinges of rose to purple   … … … … … … … … **3. P. tankervilleae**
   Stem elongate; peduncle arising from side of stem; flowers not as above
   … 2
2. Flowers pale yellow-green at base veined with diffuse red lines at base of sepals, lip pale yellow-green at base veined with red, with white hairs above; column and anther hirsute … … … … … … … … … … … … … … … **2. P. robertsii**
   Flowers white with a yellow lip; column and anther glabrous
   … … … … … … … … … … … … … … … … … … … **1. P. amboinensis** var.**papuanus**

1. **P. amboinensis** *Blume*, Mus. Bot. Lugd. Bat. 2: 180 (1852). Type: Ambon, *Blume* s.n. (holotype L).

   var. **papuanus** (*Schltr.*)*Schltr.* in Fedde, Rep. Sp. Nov., Beih. 1: 375 (1912).
   Phaius papuanus Schltr. in K. Schum. & Laut., Nachtr. Fl. Deutsch. Sudsee: 139 (1905). Type: New Guinea, *Schlechter* 14595 (holotype B).
   Phaius graeffei Reichb. f. in Seem., Fl. Vit.: 299 (1905); **synon. nov**. Type: Samoa, *Graeffe* s.n. (holotype W).

   Stem elongated. *Inflorescence* lateral from side of stem, up to 1 m long. *Flowers* white with a yellow lip; sepals and petals 2–2.5 cm long; dorsal sepal ovate; lateral sepals obliquely ovate; petals linear-obovate; lip trilobed in apical part, 3 cm long, 2 cm wide when flattened, lateral lobes oblong, rounded on front, midlobe triangular, callus of 2 fleshy ridges which are finely pubescent, spur minutely globose; column c. 2.5 mm long, glabrous. (See fig. 6).

   DISTRIBUTION: Anatom, Banks Islands (Vanua Lava), Erromango and Espiritu Santo. Also in New Guinea, Bougainville, the Solomon Islands, the Cook Islands and Fiji.
   HABITAT: Lower montane and montane ridge-top forest, in shade in leaf litter; 330–1000 m.
   COLLECTIONS: *Bernardi* 13318 (G); *Cabalion* 1732 (P, PVNH); *Cribb & Wheatley* (sight record); *Morat* 6431 (P); *Morrison* in RBG Kew 126 (K); *Wheatley* 349 (K, PVNH).

   Phaius neocaledonicus Rendle from New Caledonia is similar and may prove to be conspecific.

2. **P. robertsii** *F. Muell.* in Wing's South Sc. Rec. 3 (12): 265 (1883). Type: New Caledonia, *Layard* s.n. (holotype MEL).

Plant up to 60 cm tall. *Stem* stout, erect. *Leaves* 4–6, suberect, scattered along stem, lanceolate-elliptic, acuminate, 9–30 cm long, 2–6.5 cm wide, shortly petiolate and sheathing at base. *Inflorescence* lateral, up to 30 cm long, erect, distantly few-flowered. *Flowers* pale greenish yellow, with diffuse red lines at the base of sepals, lip pale yellow-green at base veined with red, white with white hairs above; sepals lanceolate, acuminate, 3.5–4 cm long; petals lanceolate, acute, 3.5 cm long; lip obscurely trilobed in apical part, 3 cm long, 2.5–2.8 cm wide, densely pubescent all over, lateral lobes erect, oblong, rounded, midlobe transversely oblong, rounded on front, shortly apiculate, callus of 2 obscure fleshy ridges in centre of lip and 2 short parallel rough ridges on midlobe, spur very short, conical, 1–2 mm long; column 1.8–2 cm long, densely pubescent beneath, with an erose hooded apex.

DISTRIBUTION: Banks Islands (Vanua Lava). Also in New Caledonia.
HABITAT: Terrestrial in deep to medium shade in forest, 550 m.
COLLECTION: *Wheatley* 363 (K, PVNH).

. 3. **P. tankervilleae** (*Banks ex L'Her.*) *Blume*, Mus. Bot. Lugd. Bat. 2: 177 (1856).
Limodorum tankervilleae Banks ex L'Her. in Sert. Angl.: 28 (1789). Type: Designated by Banks, cited by L'Her., from a plant from China cultivated in London (holotype K!).

For full synonymy see Hallé (1977).

Stem thickened into a short, many-jointed pseudobulb. *Inflorescence* lateral, from base of pseudobulb, up to 1.5 m tall. *Flowers* 4–12; sepals white, pale rose or orange-brown; lip white with tinges of rose to purple; sepals c. 6 cm long; dorsal sepal oblong, acuminate; lateral sepals ovate, acuminate; petals obovate, acuminate, 4–5 cm long; lip obscurely trilobed, 5 cm long, 4.5 cm wide when flattened, adnate to the base of the column, lateral lobes inrolled to form a tight tube which encompasses the column, lateral lobes obliquely broadly oblong, midlobe semicircular, apiculate, margins undulate-crisped, disc with an oblong-ovate, acuminate raised plate which extends from the base to the apex of the midlobe and is hirsute near the apex, spur conical, curved, c. 7 mm long; column dilated near the apex, c. 20 mm long, sparsely pubescent. (See fig. 6, plate 2a).

DISTRIBUTION: Anatom, Erromango, Espiritu Santo and Malekula. Widely distributed from Asia, S.E. Asia and the Malay archipelago to New Guinea, New Caledonia, the Horn Islands, Fiji, Samoa, Australia and New Zealand.
HABITAT: Amongst bracken and tree ferns, in damp conditions, 100–200 m.
COLLECTIONS: *Baker* 7 (BM); *Bernardi* 12980 (K, G); *Cheesman* 94 (K); *Hallé* in RSNH 6408 (P).

The specific epithet has been spelt in many different ways. The plant was named after the Countess of Tankerville.

Fig. 6. *Phaius amboinensis* var. *papuanus*. **A**, habit × ⅙; **B**, flower × $\frac{11}{13}$; **C**, lip × ⅞; **D**, lip × ⅞; **E**, dorsal sepal × ⅞; **F**, petal × ⅞; **G**, lateral sepal × ⅞; **H**, column × ⅞; **J**, pollinia × 3; **K**, anther cap × 3. *P. tankervilleae*. **L**, flower. A–K drawn from *Wickison* 92 (Kew spirit no. 51500); **L** from slide. All drawn by Sue Wickison.

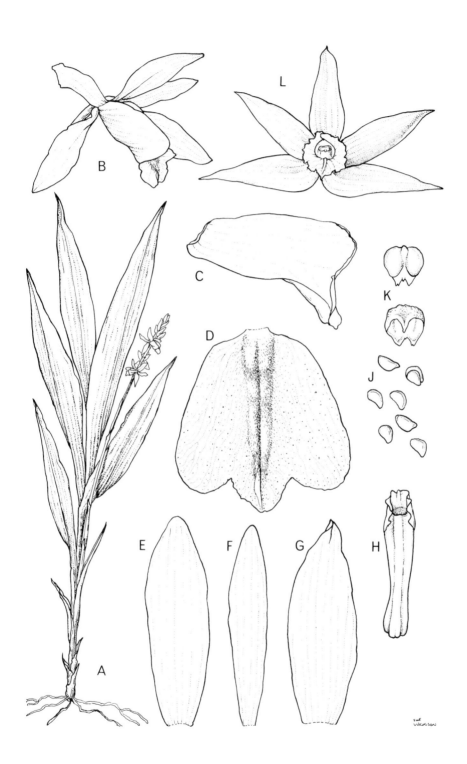

29. **SPATHOGLOTTIS** Blume

Terrestrial. Pseudobulbs small to large. *Leaves* lanceolate to broadly lanceolate, up to 1 m long, plicate, petiolate. *Inflorescence* from a basal leaf axil, tall and slender, with a few sheaths along its length; rhachis bearing a succession of many flowers. *Flowers* pink, purple or white; pedicels long and slender; sepals and petals free, about equal, spreading widely; lip trilobed, with a callus at base; column slender, curved, without a foot; pollinia 8.

A genus of about 40 species from tropical Asia to Australia and the Pacific Islands. Four species in Vanuatu.

1. Midlobe of lip lacking a claw ... **2. S. petri**
   Midlobe of lip with a claw ... 2
2. Callus at base of lip glabrous or slightly hairy ... **3. S. plicata**
   Callus at base of midlobe strongly hairy ... 3
3. Lateral lobes of lip at right angles to the axis of the lip when the lip is flattened ... **1. S. pacifica**
   Lateral lobes forming a 45–50 degree angle with the axis of the lip when the lip is flattened ... **4. S. unguiculata**

1. S. pacifica Reichb. f. in Seem., Fl. Vit.: 300 (1868). Type: Fiji, *Harvey* s.n. (holotype K!).
Limodorum unguiculatum sensu Seem., Syn. Pl. Vit., 12 (1892), non Labill. (1824). *Spathoglottis plicata* auct. non Blume.

Inflorescence up to 1 m long, pubescent; raceme dense. *Flowers* 10 or more, white to pink; pedicel c. 4 cm long including ovary, densely covered with brown hairs; sepals and petals 1.6–2.2 cm long; lip T-shaped when flattened, with the lateral lobes making a right angle with the axis of the lip and slightly curved forward in apical half, c. 6 mm long, 4 mm wide, midlobe c. 1 cm long, with the apical lamina transversely oblong with a crisped margin, 0.6–1 cm wide, calli 2, long and attenuate from the base of the claw onto the apical lamina, hairy, white to yellow. (See fig. 7, plate 2e).

DISTRIBUTION: Erromango and Espiritu Santo. Also in the Solomon Islands, the Horn Islands, Fiji, Samoa, Tahiti and Tonga.
HABITAT: Common in grassy open spaces and on sides of gorges, 400–900 m.
COLLECTIONS: *Cribb & Wheatley* 28 & 75 (K, PVNH).

2. S. petri Reichb. f. in Gard. Chron. ser. 2, 8: 392 (1877). Type: Vanuatu, *P. Veitch* s.n. (holotype W).

Inflorescence with peduncle up to 70 cm long, glabrous; raceme c. 15 cm long. *Flowers* 6–20, rose-pink to magenta, hairy on outer surface of sepals; pedicel c. 3–4 cm long including ovary, tomentose with brown hairs; sepals c. 2 cm long; lip c. 1.6 cm long, 1.6 cm wide, nearly round to oblong, trilobed, with the lateral lobes large, equal to or sometimes larger than the midlobe, rectangular and often broadest at apex, c. 1 cm long, at 45 degrees with the axis of the lip when flattened, midlobe 0.8–1.0 cm long with a very broad, transversely elliptic or circular lamina, callus rhomboid and very shaggy. (See fig. 7, plate 2f).

DISTRIBUTION: Ambae, Ambrym, Anatom, Banks Islands (Vanua Lava), Efate, Epi, Erromango, Espiritu Santo, Pentecost and Tanna. Also in New Caledonia.

HABITAT: Open grassy slopes and periodically burnt hillsides with *Pteridium* and *Gleichenia*, c. 700 m.

COLLECTIONS: *Bourdy* 116 (P) & 904 (PVNH); *Cabalion* 1357 & 1451 (PVNH); *Cheesman* 14 (K); *Green* in RSNH 1032 (K, P, PVNH), 1253 (K); *McKee* 34475 (P); *Morat* 5220 & 6033 (P); *Morrison* in RBG Kew 55 & 121; *de la Rüe* s.n. (P); *Slade* s.n. (K); *Walter* 294 (PVNH); *Wheatley* 89 (K, PVNH).

3. S. plicata *Blume* in Bijdr. Fl. Ned. Ind.: 401 (1825). Type: Java, *Blume* s.n. (holotype P!).
Spathoglottis vieillardii Reichb. f., Bijdr. Fl. Ned. Ind.: 401 (1825). Type: New Caledonia, *Vieillard* 1302 (holotype P!).
Bletia angustifolia Gaud. in Freyc., Voy. Bot.: 421 (1829). Type: Moluccas, *Freycinet* s.n. (holotype P!).
Spathoglottis unguiculata auct. non (Labill.) Reichb. f. (1868).
Spathoglottis angustifolia (Gaud.) Benth. & Hook., Gen. Pl. 3: 512 (1883).
Spathoglottis daenikeri Kraenzl. in Viertelj. Nat. Ges. Zur. 74: 80 (1929). Type: New Caledonia, *Daeniker* 1622 (holotype Z).

Inflorescence 60–100 cm long, hairy. *Flowers* 3–38, white to purple; sepals 1.6–3 cm long, hairy; lip T-shaped when flattened, lateral lobes at right angles to midlobe, oblong, 7–8 mm long, midlobe much longer than lateral lobes, 1–1.3 cm long, with a long narrow claw, dilated at the apex, callus glabrous, bipartite, with each lobe triangular–ovate and attenuated in front, yellow, in front of the calli are 2 triangular teeth which are usually slightly hairy. (See fig. 7).

DISTRIBUTION: Anatom, Erromango, Pentecost and Tanna. Widely distributed from Asia, S.E. Asia and the Malay archipelago to New Guinea, the Solomon Islands, New Caledonia, the Horn Islands, Samoa and Tonga.
VERNACULAR NAME: Evére
HABITAT: Common from the coast to inland in all the wetter areas, sea level to 500 m.
COLLECTIONS: *Cabalion* 960 (P); *Hoock* s.n. (P); *Raynal* in RSNH 16136 (P); *Wheatley* 162 (K, PVNH).

Spathoglottis plicata is the most widespread and widely cultivated species of this genus and consequently has been described many times, several varieties have been selected, including white ones. It has been introduced into the Hawaiian Islands and elsewhere in S.E. Asia, Africa and the Americas.

4. S. unguiculata (*Labill.*) *Reichb. f.* in Seem., Fl. Vit.: 300 (1868).
Limodorum unguiculatum Labill., Sert. Austr.-Caled.: 19, t.25 (1824). Type: New Caledonia, *J. Labillardiere* s.n. (holotype P!).
Spathoglottis deplanchei Reichb. f. in Linnaea 41: 86 (1877). Type: New Caledonia, *Deplanche* 163 (holotype P!).
Spathoglottis breviscapa Schltr. in Engler, Bot. Jahrb. 39: 65 (1906). Types: New Caledonia, *Schlechter* 15486 (holotype B; isotypes BR & K!).
Spathoglottis schinziana Kraenzl. in Viertelji. Naturf. Ges. Zurich, 74: 80 (1929). Type: Loyalty Islands, *A. V. Daeniker* 2492 (holotype Z).

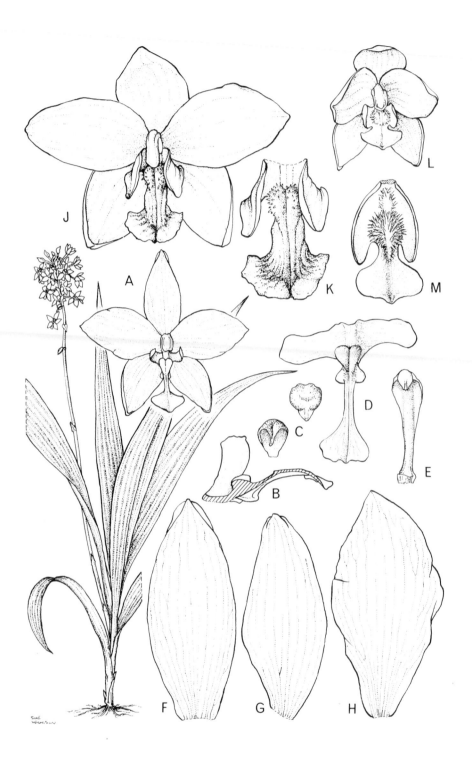

Inflorescence with peduncle up to 80 cm long, glabrous; raceme c. 15 cm long. *Flowers* 10 or more, rose to dark purple, hairy on outer surface of sepals and pedicel; pedicels 3–4 cm long including ovary; sepals and petals c. 2 cm long; lip c. 1.6 cm long, 1.4 cm wide, nearly round to oblong, trilobed, lateral lobes large, equal or sometimes larger than the midlobe, rectangular and often broadest at the apex, c. 1 cm long, at c. 30 degrees to the axis of the lip when flattened, midlobe with a short claw and a broad, transversely elliptic or circular apical lamina, callus bilobed, each lobe broadly ovoid, hairy.

DISTRIBUTION: Ambrym, Anatom, Banks Islands, Erromango and Tanna. Also in New Caledonia, the Isle of Pines, the Horn Islands and Fiji.

HABITAT: In grassland, along paths in forest and groves and on periodically burnt hillsides, 180–650 m.

COLLECTIONS: *Bernardi* 12931 (K, P) & 13176 (P); *Cabalion* 598 (P); *Cheesman* in RBG Kew 10 (K); *Hoock* s.n. (P); *Kajewski* 306 (K); *Morat* 6076 (P); *de la Rüe* s.n. (P); *Veillon* 3022 (P); *Wheatley* 362 (K, PVNH).

30. **COELOGYNE** Lindley

Epiphytic. Pseudobulbs 1- or 2-leaved. *Leaves* elliptic. *Inflorescence* terminal, erect or pendulous, with few or many large or fairy large flowers. *Flowers* with sepals often concave; petals narrower than sepals; lip trilobed, the lateral lobes widening gradually from the base of the lip and erect, midlobe of lip with longitudinal keels which are often papillose or verrucose; column long, winged around the top; pollinia 4.

A large genus of about 300 species from tropical Asia to the Pacific Islands. Two species in Vanuatu.

Pseudobulbs 2-leaved; leaves lanceolate, elliptic, or oblanceolate, 17–40 cm long, 2.5–11.5 cm wide   ... ... **2. C. macdonaldii**
Pseudobulbs 1-leaved; leaves broadly elliptic to elliptic-lanceolate, 18–35 cm long, 8–12 cm wide   ... ... **1. C. lycastoides**

1. C. lycastoides *F. Muell. & Kraenzl.* in Oest. Bot. Zeit. 45: 179 (1895). Type: Samoa, *Betche* 24 (holotype MEL).
Coelogyne whitmeei Schltr. in Fedde, Rep. Sp. Nov. 11: 41 (1912); **synon. nov.** Type: Samoa, *Rev. Whitmee* s.n. (holotype B);

Pseudobulb ovoid to cylindric, 4–8 cm long, 0.9–3 cm wide, pale green. *Leaf* solitary, elliptic to elliptic-lanceolate, acute, 18–35 cm long, 8–12 cm wide, mid-green, petiolate. *Inflorescence* with an arcuate peduncle and zig-zag rhachis; bracts deciduous, linear, 4–6 cm long. *Flowers* greenish-yellow with orange-brown on lip but apex of lip and column white to yellow; dorsal sepal oblong-elliptic, obtuse, 4.2 cm long; lateral sepals obliquely oblong-elliptic, obtuse, 4.2

Fig. 7. *Spathoglottis plicata*. **A**, habit and flower life size; **B**, longitudinal section of lip × 2; **C**, anther cap × 4; **D**, lip × 2; **E**, column × 2; **F**, dorsal sepal × 2; **G**, lateral sepal × 2; **H**, petal × 2. *S. pacifica*. **J**, flower × 2; **K**, lip × 3; *S. petri*. **L**, flower × 2; **M**, lip × 3. **A** drawn from slide; **B–H** from *Hunt* 2939 (Kew spirit no. 28575); **J–K** from *Simmonds* s.n. (Kew spirit no. 18343); **L** from slide; **M** from *Lecoufle* s.n. (Kew spirit no. 36456). All drawn by Sue Wickison.

cm long; petals linear, obtuse, 4.5 cm long; lip trilobed, 4 cm long, 2.6 mm wide, lateral lobes narrowly elliptic, rounded in front, midlobe oblong, obtuse, with callus of 4–6 verrucose lines in central part of lip; column 2.5–2.8 cm long. *Fruits* 6-winged, 4–5 cm long, c. 3 cm wide.

DISTRIBUTION: Ambae, Anatom, Espiritu Santo, Malekula and Pentecost. Also in New Caledonia and Samoa.

HABITAT: Mixed evergreen forest, sea level to 700 m.

COLLECTIONS: *Cabalion* 451 (P); *Hallé* in RSNH 6420 (K, P); *Hoock* s.n. (P); *McKee* 32188 (P); *Morat* 5421 & 6442 (P); *Morrison* in RBG Kew 146 (K); *Suprin* 360 (P); *Wheatley* 77 (K, PVNH).

2. C. macdonaldii *F. Muell. & Kraenzl.* in Oest. Bot. Zeit. 44: 209 (1894). Type: Vanuatu, *MacDonald* s.n. (holotype MEL).
Coelogyne lamellata Rolfe in Kew Bull. 1895: 36 (1895); **synon. nov.** Type: Vanuatu, ex *Sander* (holotype K!).

Pseudobulb ovoid to narrowly ovoid, 2.5–6.5 cm long, 0.8–4 cm wide, pale green. *Leaves* 2, lanceolate, elliptic or oblanceolate, acuminate, 17–40 cm long, 2.5–11.5 cm wide; shortly petiolate. *Inflorescence* produced from between developing leaves of new shoot, 8–30 cm long; peduncle arcuate, with flowers produced in succession, usually 1 at a time; rhachis zig-zag; bracts deciduous, linear, acuminate, 3–5 cm long. *Flowers* pale green with a red-brown callus on lip; dorsal sepal narrowly elliptic, acute, 3.5–4.5 cm long; lateral sepals obliquely oblong, obtuse, 3–4 cm long; petals linear, acute, 3.5–4 cm long; lip trilobed, 2.5–3.3 cm long, 1.8–2 cm wide, lateral lobes elliptic, rounded in front, callus of 7–9 verrucose lines, from base of lip to centre of midlobe; column 1.8–2.2 cm long. *Fruits* 6-winged, up to 7 cm long, 2.5 cm wide. (See plate 3e).

DISTRIBUTION: Banks Islands (Vanua Lava), Erromango, Espiritu Santo, Malekula and Pentecost. Also in Fiji.

HABITAT: Ridge-top forest and lower montane forest, 270–1100 m.

COLLECTIONS: *Baker* 11 (BM); *Bourdy* 875 (PVNH); *Cabalion* 1750 (PVNH); *Cribb & Wheatley* 48 (K, PVNH); *Morat* 7463 (P); *Morrison* in RBG Kew 145 (K); *Raynal* in RSNH 15988 (P) & 16323 (K, P); *de la Rüe* s.n. (P); *Sander* s.n. (K); *Veillon* 2430 & 5513 (P); *Walter* 240 (PVNH); *Wheatley* 108 & 410 (K, PVNH).

31. **PHOLIDOTA** Lindley ex Hooker

Epiphytic or lithophytic. *Pseudobulbs* close together, ovoid-conical, sturdy and usually rather swollen. *Leaf* solitary from apex of pseudobulb. *Inflorescence* terminal, from apex of pseudobulb, pendent, slender, bearing many small flowers alternately in 2 ranks; the subtending bracts large and concave, persistent. *Flowers* small, fleshy; sepals concave; lip bipartite, with a saccate base and small, deflexed apical blade; column short with a wide hood around the anther; pollinia 2.

A genus of about 20 species from tropical Africa to Australia and the Pacific Islands. A single species in Vanuatu.

P. imbricata W.J. Hooker, Exotic Fl. 2: 138 (1825). Type: Himalayas, *Wallich* s.n., cult. Kew (1824) (probably not preserved); lectotype: the drawing and description in Exotic Fl. 2: 138, chosen by Seidenf., Opera Botanica 89: 102 (1986).

Pholidota grandis Kraenzl. in Bull. Soc. Bot. Fr. 76: 301 (1929). Type: Vanuatu, Efate, *Levat* s.n. (holotype MPU) non Ridley (1908).

Pholidota spectabilis Kraenzl. ex Guillaumin in Ann. Mus. Col. Marseille, ser. 2, 15 (1948). Type: as for *P. grandis*.

For full synonymy see de Vogel (1988).

Plant 10–66 cm high. *Pseudobulbs* 2–11 cm long, midgreen to dull grey-green. *Leaf* oblong or obovate-oblong to linear, 17–52 cm long, 1.6–9 cm wide, thickly coriaceous, midgreen to dark green above, lighter green below; petiole 0.5–9 cm long. *Inflorescence* with a terete peduncle, up to 46 cm long; rhachis up to 41 cm long, pendulous. *Flowers* 14–126, white to cream-coloured, may be tinged yellowish; pedicel 2–5 mm long; dorsal sepal and petals forming a hood; dorsal sepal ovate, 5–6 mm long; lateral sepals ovate, 5–7 mm long; petals falcate, 4.5–6 mm long; lip midlobe trilobed, 4 mm long, 5 mm wide, lateral lobes erect, triangular to almost semi-circular, apex bilobed, saccate base of lip with 3 yellow calli. (See plate 3f).

DISTRIBUTION: Anatom, Banks Islands (Vanua Lava), Efate, Erromango, Maewo, Malekula, Pentecost and Tanna. Widely distributed from Asia, S.E Asia and the Malay archipelago to New Guinea, the Solomon Islands, the Santa Cruz Islands, the Loyalty Islands, New Caledonia, Fiji and Australia.

HABITAT: Mangrove forest, strand forest, marsh forest, lowland and hill evergreen forest, dry deciduous forest to lower montane forest, sea level to 900 m.

COLLECTIONS: *Bourdy* 535 (PVNH); *Cabalion* 1009, 1596, 1597, 1669, 1745 & 2563 (PVNH); *Cribb* 1773 (K); *Hallé* in RSNH 6414 (K, P); *Hoock* s.n. (P); *Morrison* in RBG Kew 81 (K); *Wheatley* 172 & 360 (K).

USES: In Vanuatu the seeds of *P. imbricata* have been used as a substitute for face powder (de Vogel, 1988).

There has been confusion in the past between *Pholidota imbricata* and *P. pallida*. De Vogel (1988) has established that the correct application of the name *P. pallida* is to a species with a limited distributiom in India.

32. **LIPARIS** L.C. Richard

Epiphytic, lithophytic or *terrestrial* plants. *Leaves* 1–3, narrowly lanceolate or ovate. *Inflorescence* terminal, racemose and few–many-flowered. *Flowers* small to medium-sized, may be self-fertilizing, green, orange or purple; sepals and petals free, often reflexed and narrow; lip variable; column footless, rather elongated, curved or bent at apex, may be swollen at base; pollinia 4.

A cosmopolitan genus of about 250 species. Six species are recorded from Vanuatu, *Liparis aaronii* and *L. pullei* being new records.

1. Leaves broadly ovate, 4–8.5 cm wide; flowers maroon with a green column
   ... ... **5. L. layardii**
   Leaves narrowly lanceolate, less than 3.5 cm wide; flowers not maroon
   2

2. Bracts distichous and imbricate ... ... .j. **4. L. gibbosa**
   Bracts not imbricate 3
3. Pseudobulbs elongated, 7–24 cm long, swollen at base; leaves 2; lip orange,
   red or red-brown ... ... **3. L. condylobulbon**
   Pseudobulbs compact, up to 4 cm high, not swollen at base; leaf solitary;
   lip green 4
4. Leaf 7.5–10 cm long, 0.7–1.1 cm wide; lip oblong and entire
   ... ... **2. L. caespitosa**
   Leaf 15–25 cm long, 1.7–3.0 cm wide; lip not oblong 5
5. Lip semicircular ... ... **6. L. pullei**
   Lip oblong, deeply 2-lobed at apex, with toothed lobes ... ...**1. L. aaronii**

1. L. aaronii *Cribb & B. Lewis* in Orchid Rev. in press (1989). Type: Vanuatu, Espiritu Santo, *Cribb & Wheatley* 105 (holotype K!).

Epiphytic. Pseudobulbs 1-leaved, clustered, conical, 1–2 cm high, 5 mm wide, covered in cataphylls. *Leaf* erect, lanceolate, tapering basally, 15–24 cm long, 1.5–3.0 cm wide, mid green, with 3–4 prominent longitudinal veins. *Inflorescence* an erect raceme, 10–15 cm long, 5–10-flowered; bracts lanceolate, acuminate, 3–6 mm long, green. *Flowers* green or lime-green; pedicel and ovary 8–10 mm long; ovary prominently keeled; dorsal sepal oblong, obtuse, 4.5–5.5 mm long, 2–2.5 mm wide; lateral sepals free, obliquely oblong, obtuse, 4.5–6 mm long, 3 mm wide; petals spreading or reflexed, linear, rounded at apex, 4–5 mm long, 0.5–0.8 mm wide; lip oblong, deeply bilobed at apex, 6–7 mm long, 3 mm wide, with a short tooth in the apical sinus, apical lobes 2–4 denticulate; column swollen at base, incurved, 2 mm long, 2.5 mm wide. (See fig. 8).

DISTRIBUTION: Ambae, Espiritu Santo and Tanna.
HABITAT: *Metrosideros* and montane forest, 1300–1800 m.
COLLECTIONS: *Cribb & Wheatley* 105 (K, PVNH); *Morat* 5902 & 6034 (P); *Chew Wee-Lek* in RSNH 245 (K); *Wheatley* 46 (K).

Liparis aaronii is closely allied to the New Guinea species, *L. werneri* Schltr. but differs in having narrower leaves, smaller flowers and a lip that is shorter and has much shorter apical lobes.

2. L. caespitosa *(Thouars) Lindley* in Bot. Reg. 11: t.882 (1825).
Malaxis caespitosa Thouars, Orch. Iles Austr. Afr.: t.90 (1822). Type: Mauritius, Thouars s.n. (holotype P).
Liparis neoguineensis Schltr. in Fedde, Rep. Sp. Nov., Beih. 1: 209, fig. 289 (1911).
Type: New Guinea, *Schlechter* 13934 (holotype K!).
For full list of synonymy see Seidenfaden (1976).

Epiphytic or lithophytic. *Pseudobulbs* c. 1.0 cm high, 3 mm wide. *Leaf* solitary, lanceolate, 7.5–10 cm long, 0.7–1.1 cm wide, tapering at base to 1–2 mm wide, light green. *Inflorescence* 8–9 cm long. *Flowers* numerous (8–25), usually self-pollinating; yellow to light green; pedicels c. 4 mm long; sepals lanceolate,

Fig. 8. *Liparis aaronii*. **A**, habit × ⅔; **B**, flower × 4; **C**, column × 6; **D**, lip × 6; **E**, dorsal sepal × 6; **F**, petal × 6; **G**, lateral sepal × 6. Drawn from *Cribb & Wheatley* 105 (Kew spirit no. 53177) by Sue Wickison.

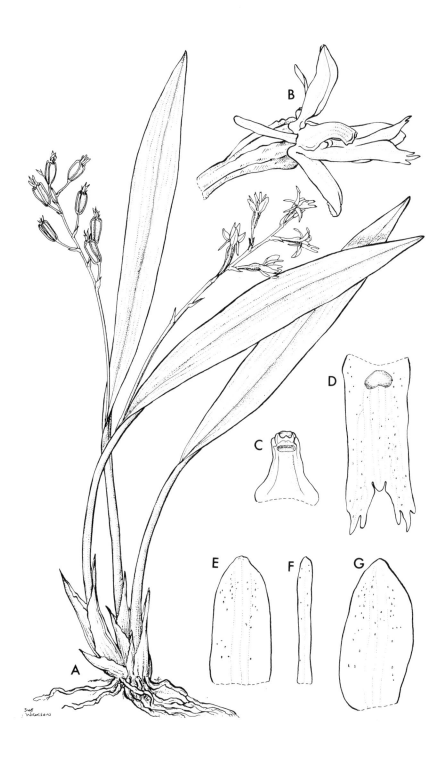

reflexed, c. 2.2 mm long; petals linear, reflexed, c. 2.2 mm long; lip decurved, fleshy, oblong when flattened, c. 2.5 mm long, 1.5 mm wide; column 1.2 mm long. (See fig. 9).

DISTRIBUTION: Erromango. Widely distributed from Madagascar and Africa to Asia, S.E. Asia, the Malay archipelago, New Guinea, the Solomon Islands and Fiji.

HABITAT: Rain forest, 550–670 m.

COLLECTIONS: *Green* in RSNH 1350 (K); *Raynal* in RSNH 16242 (K, P).

3. L. condylobulbon *Reichb. f.* in Hamb. Gartenz. 18: 34 (1862). Type: Moulmein (Burma), *Parish* number questionable (material probably lost. A sketch in Reichbenbach's herbarium in Vienna is questioned by J.J. Smith (1927) as representing the type).

Liparis longipes auct. non Lindley, Gen. Sp. Orch. Pl.: 30 (1830).

Liparis nesophila Reichb. f. in Otia Bot. Hamb. 1: 56 (1878). Type: Fiji, *Seemann* 614 (holotype K!).

Liparis confusa J.J. Smith, Orch. Java: 275 (1905). Type: Java, *J.J. Smith* 903 (holotype BO).

Liparis elegans sensu Ames in Journ. Arn. Arb. 14: 105 (1933), non Lindley.

Epiphytic, forming large clumps. *Pseudobulbs* elongated, up to 24 cm long, swollen at base, 1.2–2 cm wide. *Leaves*, usually 2, lanceolate, 12–20 cm long, 1.3–2.5 cm wide. *Inflorescence* 6–22 cm long. *Flowers* numerous, 15–25, usually self-pollinating, pale green, scented; pedicels c. 6 mm long; sepals lanceolate, reflexed, c. 2.5 mm long; petals linear, reflexed, c. 2 mm long; lip decurved, oval when flattened, bifid at apex, with margin of basal half minutely ciliate, orange, red or red-brown; column c. 1.5 mm long. (See fig. 9, plate 3c).

DISTRIBUTION: Anatom, Banks Islands, Efate, Erromango, Espiritu Santo, Malekula, Pentecost and Tanna. Widely distributed from Asia and S.E. Asia to the Malay archipelago, New Guinea, the Solomon Islands, the Horn Islands, Fiji and Samoa.

VERNACULAR NAME: Ute-melme-pohl.

HABITAT: Wide ranging from lowland *Kleinhovia-Dendrocnide* forest on alluvial flats to montane forest, and secondary forest, usually in shade, often found in gullies, 20–500 m.

COLLECTIONS: *Bernardi* 13090 (P); *Bourdy* 858 (P, PVNH); *Bregulla* 26 (K); *Cabalion* 253 (P), 1106, 1631, 1974 (PVNH) & 2483 (P, PVNH); *Cribb* 52 (K); *Cribb & A. Morrison* 1792 (K); *Hallé* 13090 (K), in RSNH 270, 4534 (K), 6394 (K, P), 6368 (P) & 6395 (PVNH); *Kajewski* 455 (K); *Morat* 5612 & 6078 (P); *Morrison* in RBG Kew 87, 88, 89, 90, 91 & 92 (K); *Raynal* in RSNH 15997 (P) & 16380 (PVNH); *Veillon* in RSNH 3960 (P) & 4534 (K, P, PVNH); *Chew Wee-Lek* in RSNH 270 (K, P, PVNH); *Wheatley* 191 (K, PVNH).

Fig. 9. *Liparis condylobulbon*. **A**, habit × ⅔; **B**, lip × 10; **C**, flower × 10; **D**, dorsal sepal × 10; **E**, lateral sepal × 10; **F**, petal × 10; **G**, column × 14; **H**, anther cap × 14; **J**, pollinia × 14. *L. caespitosa*. **K**, flower × 10; **L**, pollinia × 14; **M**, anther cap × 14; **N**, lip × 10; **O**, column × 14; **P**, dorsal sepal × 10; **Q**, lateral sepal × 10; **R**, petal × 10; **S**, habit × ⅔. **A** drawn from *Hallé* 6394; **B–J** from *Wickison* 50 (Kew spirit no. 50871); **K–R** from *Cribb et al* 5043 (Kew spirit no. 48649); **S**, from *Ridsdale & Lavarack* 30674. All drawn by Sue Wickison.

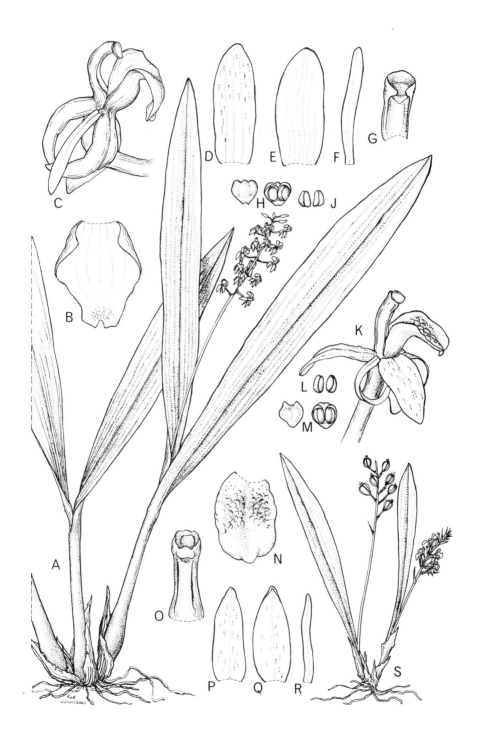

4. L. gibbosa *Finet* in Bull. Soc. Bot. Fr. 55: 342, fig. 32–44 (1908). Type: Java, *Blume* s.n. (holotype P!).
Liparis disticha auct. non (Thouars) Lindley in Bot. Reg. 11: t.882 (1825).

Epiphytic or lithophytic, growing in clumps. *Pseudobulbs* up to 4 cm high, 1.2 cm wide, pale green, may be sheathed in cataphylls. *Leaf* usually solitary, lanceolate, 11–15 cm long, 0.7–1.6 cm wide, pale green to rich green. *Inflorescence* with peduncle 10–20 cm long, orange-green; raceme 2–3 cm long; bracts distichous and imbricate, green to orange-brown. *Flowers* usually opening 1 at a time, bright orange to dark red to brown; pedicels c. 1 cm long; sepals lanceolate, reflexed, c. 5 mm long; lip decurved, oblong when flattened, tapering towards apex, c. 4.5 mm long, 2 mm wide, with a basal callus; column swollen at base, c. 2 mm long, 2.5 mm wide, pale green. (See fig. 10).

DISTRIBUTION: Ambae, Efate, Espiritu Santo and Pentecost. Widely distibuted from S.E. Asia, the Malay archipelago to New Guinea, the Solomon Islands, New Caledonia, Fiji and Samoa.
HABITAT: *Metrosideros-Weinmannia* forest and montane ridge-top forest, in shade, 350–1600 m.
COLLECTIONS: *Bourdy* 726 (P); *Cabalion* 1042 (PVNH); *Hallé* in RSNH 16380 (PVNH); *Raynal* in RSNH 16380 (K, P); *Chew Wee-Lek* in RSNH 208 (K, P); *Veillon* 3985 (P); *Wheatley* 73 & 111 (K).

5. L. layardii *F. Muell.* in Wing's South Sci. Rec.: 1 (1885). Type: New Caledonia, *Layard* s.n. (holtype MEL).
Liparis stricta Schltr. in Fedde, Rep. Sp. Nov. 9: 95 (1910). Type: Samoa, *Schlechter* 134 (isotype K!), non J.J. Smith (1907).
Liparis mataanensis J.J. Smith in Bull. Jard. Bot. Buitenz. 8: 56 (1912). Type: as for *L. stricta*.

Terrestrial. Stem up to 25 cm high, sheathed in cataphylls. *Leaves* 2–3, ovate, 10–13 cm long, 4–8.5 cm wide, glossy, mid green, paler beneath. *Inflorescence* 18–30 cm long, purple. *Flowers* 6–12, purple; pedicels c. 2 cm long, purple; sepals lanceolate, reflexed; dorsal sepal 12.5–14 mm long, 1 mm wide; lateral sepals 11 mm long, 1 mm wide; petals linear, reflexed, 12 mm long, 0.7 mm wide; lip ovate, c. 10 mm long, 8–9 mm wide; column c. 5 mm long, green. (See fig. 10, plate 3d).

DISTRIBUTION: Ambae, Anatom, Erromango and Espiritu Santo. Also in the Solomon Islands, New Caledonia and Samoa.
HABITAT: Dense rain forest to mossy montane forest, 380–1600 m.
COLLECTIONS: *Cabalion* 2359 (P); *Cribb & Wheatley* 52 & 85 (K, PVNH); *Green* in RSNH 1128 (K) & 1159 (K, P); *Morrison* in RBG Kew 133 & 136 (K); *Raynal* in RSNH 16382 (K, P, PVNH); *Wheatley* 32 & 35 (K, PVNH).

6. L. pullei *J.J.Smith* in Bull. Jard. Bot. Buitenz. ser. 2., 13: 56 (1914). Type: New Guinea, *A. Pulle* 272 (holotype BO).

Fig. 10. *Liparis gibbosa*. **A**, habit × ⅔; **B**, flower × 6; **C**, petal × 8; **D**, dorsal sepal × 8; **E**, lateral sepal × 8; **F**, lip × 8; **G**, column × 10; **H**, anther cap × 10; **J**, pollinia × 10. *L. layardii*. **K**, habit × ⅔; **L**, flower × 3; **M**, lip × 3; **N**, column × 3; **O**, lateral sepal × 3; **P**, dorsal sepal × 3; **Q**, petal × 3; **R**, anther cap × 8; **S**, pollinia × 8. A–J drawn from *Wickison* 105A (Kew spirit no. 52080); K from *Mackee* 19; L–S from *Mitchell* 5 (Kew spirit no. 50787). All drawn by Sue Wickison.

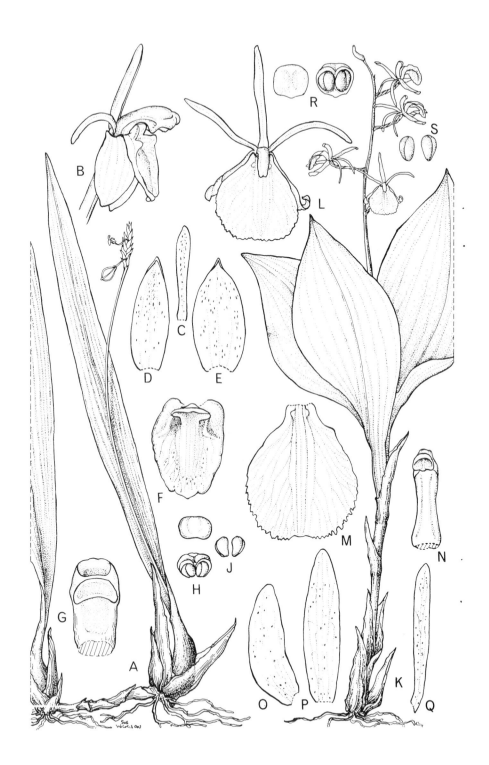

Terrestrial or epiphytic on the base of trunks. *Stem* 4–5 cm long, may be covered in cataphylls. *Leaf* solitary, up to 20 cm long, 2–3 cm wide, with 3–4 prominent longitudinal veins. *Inflorescence* 13–22 cm long. *Flowers* c. 6; pedicels c. 1.7 cm long; sepals lanceolate, reflexed, c. 1 cm long; petals linear, reflexed, c. 1 cm long; lip semicircular, 1.0 cm long, 1.1 cm wide, with a small fleshy basal callus; column 4.5 mm long.

DISTRIBUTION: Tanna. Also in New Guinea.
HABITAT: Rain forest. No information from Vanuatu, in New Guinea it grows in rain forest, at 100 m.
COLLECTION: *Morat* 6034 (NOU).

This identification is provisional as the material cited is in rather poor condition.

33. **MALAXIS** Solereder ex Swartz

Terrestrial or rarely lithophytic. *Rhizome* creeping. *Stem* erect, leafy. *Leaves* plicate, broad, ovate, acute, thin-textured, petiolate, with petioles sheathing. *Inflorescence* terminal, erect, with many small flowers. *Flowers* non-resupinate, may be self-pollinating; sepals and petals free; petals usually narrower than the sepals; lip often with a hollow near the base, which may contain nectar, usually with 2 auricles close to the sides of the column, apex often toothed; column very short.

A cosmopolitan genus of about 300 species. Four species in Vanuatu, *Malaxis dryadum* being a new record. The small flowers of this genus make their identification difficult.

1. Lip 4.5–7 mm wide, very obscurely toothed ... ...**1. M. dryadum**
   Lip less than 4.5 mm wide, with prominent teeth 2
2. Lip horse-shoe-shaped, with 4–6 teeth ... ... **4. M. xanthochila**
   Lip with rounded auricles and elongated midlobe, with 2–4 teeth 3
3. Leaves 3–4; flowers green, yellow or purple; lip midlobe entire
   ... ... **3. M. taurina**
   Leaves 7–11; flowers yellow-orange to orange; lip midlobe emarginate
   ... ...**2. M. lunata**

1. M. dryadum *(Schltr.) P.F. Hunt* in Kew Bull.: 79 (1970).
Microstylis dryadum Schltr. in K. Schum. & Laut., Nachtr. Fl. Deutsch. Sudsee: 98 (1905). Type: New Guinea, *Schlechter* 14048 (holotype B).

Stem covered by leaf bases, c. 10 cm long. *Leaves* 3–4, suberect, ovate-lanceolate, acuminate, 7 cm long, 2.4–2.9 cm wide; petiole entirely sheathing stem. *Inflorescence* c. 14 cm long; bracts linear-lanceolate, acuminate, 7–8 cm long. *Flowers* yellow-green turning yellow or dull plum-coloured; pedicels 7–10 mm long; dorsal sepal oblong-elliptic, obtuse, 4–5 mm long; lateral sepals broadly obliquely-ovate-elliptic, obtuse, 3.8–4 mm long; petals oblong, obtuse 3.8–4 mm long; lip broadly reniform, obscurely and very shortly toothed in front, 4–4.5 mm long, 4.5–7 mm wide, with auricles 4 mm long and rounded at apex, midlobe very obscurely broadly triangular, obtuse. (See fig. 11).

DISTRIBUTION: Espiritu Santo. Also in New Guinea.
HABITAT: Montane forest, 1200 m.
COLLECTIONS: *Raynal* in RSNH 16373 (K, P); *Robinson* K148 (K).

The Vanuatu specimens are close to *Malaxis dryadum* Schltr. but more material is needed before this identification can be confirmed.

2. M. lunata *(Schltr.) Ames* in Journ. Arn. Arb. 13: 129 (1932).
Microstylis lunata Schltr. in Fedde, Rep. Sp. Nov. 4: 162 (1911). Type: Vanuatu, Anatom, *Morrison* in RBG Kew 82 (holotype K!).

Stem up to 10–15 cm long, fleshy, not completely covered by leaf bases, dull purple. *Leaves* 7–11, elliptic-ovate, acuminate, 8–15 cm long, 2.5–6 cm wide, oblique at base of lamina; petioles 3.5–4 (5) cm long, sheathing at base. *Inflorescence* 15–24 cm long; peduncle and rhachis rigid; bracts reflexed, linear, acuminate, 4–6 mm long. *Flowers* yellow-orange to orange; pedicels 2.5–3 mm long; dorsal sepal oblong, obtuse to elliptic, 3–5 mm long; lateral sepals oblong-ovate, obtuse, 2.7–4.5 mm long; petals oblong to oblanceolate, obtuse, 2.5–3 mm long; lip transversely reniform, 3–4 mm long, 3.4–5 mm wide, with auricles rounded at base, the front margin with 1–2 short teeth on either side of midlobe, midlobe longer than teeth, oblong, emarginate. (See fig. 11).

DISTRIBUTION: Ambrym, Anatom, Efate, Erromango and Tanna. Also in the Solomon Islands.
HABITAT: Forest, in shade, up to 400 m.
COLLECTIONS: *Bernardi* 12996 (P); *Cheesman* s.n., A31, A74a & A75 (BM); *Green* in RSNH 1122 (K, P); *Hallé* in RSNH 6389 (P); *Quaife* 212 (K); *Morrison* in RBG Kew 82, 83, 85, 86, 132, 135, 137 (K); *de la Rüe* s.n. (P).

3. M. taurina *(Reichb. f.) Kuntze*, Rev. Gen. Pl. 2: 673 (1891).
Microstylis taurina Reichb. f. in Linnaea 41: 97 (1877). Types: New Caledonia, *Deplanche* 354 & 147 (syntypes P!).
Goodyera glabra Kraenzl. in Viertelj. Nat. Ges. Zur. 74: 69 (1929). Type: New Caledonia, *Daeniker* 2905 (holotype Z).
Microstylis procera Kraenzl., l.c.: 71. Type: New Caledonia, *Daeniker* 2905 (holotype Z).

Stem short, fleshy, c. 3 cm long, covered in leaf bases. *Leaves* 3–4, ovate-elliptic or elliptic-lanceolate, acute, 3.5–5 cm long, 1–2.5 cm wide; petioles 2–3 cm long. *Inflorescence* 8–9 (14) cm long; bracts reflexed, linear, acuminate, 3–4 mm long. *Flowers* self-pollinating, green, yellow or purple; dorsal sepal ovate, c. 5 mm long; lateral sepals oblong-ovate, 5 mm long; petals obovate, 4.5 mm long; lip 4.4 mm long, 4.4 mm wide, with rounded auricles at base, and the margin with 1–2 short teeth on either side of fleshy, entire midlobe which is longer than the teeth. (See fig. 11).

DISTRIBUTION: Anatom, Banks Islands (Vanua Lava) and Erromango. Also in New Caledonia and possibly the Solomon Islands (see note).
HABITAT: Montane forest, 700–800 m.
COLLECTIONS: *Morrison* in RBG Kew 99 & 134 (K); *Veillon* 5547 (NOU).

A sterile specimen, *Mitchell* 72, from Ranongga, Solomon Islands, is similar in habit.

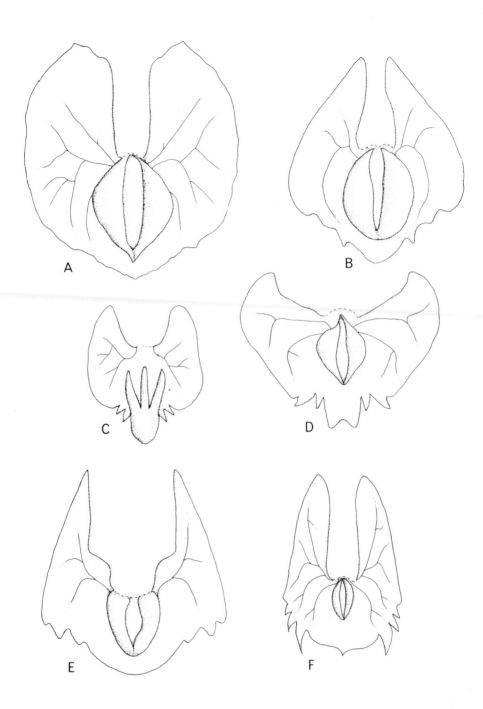

70

4. M. xanthochila *(Schltr.) Ames & C. Schweinf.* in Orch. 6: 73 (1920).
Microstylis xanthochila Schltr. in K. Schum. & Laut., Nachtr. Fl. Deutsch. Sudsee: 102 (1905). Type: New Guinea, *Schlechter* 13678 (holotype B).
Microstylis sordida J.J. Smith in Bull. Dep. Agric. Ind. Neerl. 19: 30 (1908). Type: New Guinea, *Versteeg* 1333 (holotype BO).
Malaxis neo-ebudica Ames in Journ. Arn. Arb. 13: 128 (1932); **synon. nov.** Type: Vanuatu, Tanna, *S.F. Kajewski* 137 (holotype AMES; isotype K!).

Stem fleshy, 20–30 cm long. *Leaves* 5–7, suberect, ovate, acuminate, 5.5–10 cm long, 2–3.7 cm wide; petiole 2.5–5 cm. *Inflorescence* 11–15 cm long. *Flowers* yellow, greenish buff, or purple, can be self-pollinating; pedicels c. 5 mm long; dorsal sepal elliptic-oblong, obtuse, 3–3.8 mm long; lateral sepals obliquely elliptic-ovate, obtuse; petals linear, obtuse, 3–4 mm long; lip transversely reniform, 4–4.5 mm long, 3–3.5 mm wide, basal auricles rounded, shortly 4–6 toothed on front margin, teeth shorter than small, semi-elliptic, obtuse or shortly emarginate midlobe. (See fig. 11).

DISTRIBUTION: Ambae, Anatom, Banks Islands (Vanua Lava), Efate, Erromango, Espiritu Santo, Malekula, Pentecost and Tanna. Also in New Guinea, the Solomon Islands, Fiji and Australia.
HABITAT: Forest and bush, sea level to 800 m.
COLLECTIONS: *Bernardi* 13358 (P); *Bourdy* 110 (NOU, P); *Cabalion* 1940 (K, P); *Cheesman* in RBG Kew 11 (K); *Kajewski* 837 (K); *Morat* 5242 (NOU, P) & 5478 (P); *Morrison* in RBG Kew 84 (K); *Raynal* in RSNH 16201 (K); *de la Rüe* s.n. (P); *Robinson* K154 (K); *Wheatley* 110, 208, 338 & 348 (K, PVNH).

Malaxis quaifei Rolfe, based on a specimen, *Quaife* 3 (K) from Vanuatu, is an unpublished name which appears to be identical to *M. xanthochila*, but which has purple flowers.

34. **OBERONIA** Lindley

Epiphytic. Leaves bilaterally flattened, either arranged distichously on stem or grouped in a fan at base. *Inflorescence* terminal, densely many-flowered. *Flowers* numerous and small, arranged in whorls on short pedicels, white, pale green, yellow, orange, red or brown; pollinia 4.
A genus of about 300 species from the Old World Tropics. Four species in Vanuatu.

1. Leaves 14–35 cm long; lip entire   ... ... **2. O. heliophila**
   Leaves less than 8 cm long; lip trilobed 2
2. Leaves grouped at the base in a fan   ... ... **4. O. titania**
   Leaves 2-ranked on 8–30 cm long stem   3
3. Leaves 1–2 cm long, closely appressed to stem   ... ... **3. O. imbricata**
   Leaves 2–5 cm long, spreading   ... ... **1. O. equitans**

Fig. 11. Comparative lips of *Malaxis*. **A**, *M. dryadum* × 13; **B**, *M. xanthochila* × 14; **C**, *M. taurina* × 14; **D**, *M. lunata* × 14; **E & F**, *M. xanthochila* × 14. **A** drawn from *Robinson* 148; **B** from *Robinson* K145; **C** from *Veillon* 5547; **D** from *Morrison* 85; **E** from *Cribb & A. Morrison* 1811; **F** from *Kajewski* 137. All drawn by Sarah Robbins.

1. O. equitans *(G. Forst.) Mutel* in Premier Mém. sur les Orch. Paris: 8 (1838).
Epidendrum equitans G. Forst., Fl. Ins. Austr. Prodr.: 60. (1786). Type: Tahiti, *G. Forster* 170 (holotype BM!, isotype P!).
Oberonia glandulosa Lindley, Fol. Orch. Oberonia 1: 6 (1859). Type: Pacific Islands, *Matthews* 158 (holotype K, not traced).
Oberonia flexuosa Schltr. in Engler, Bot. Jahrb. 39: 61 (1906). Type: New Caledonia, *Schlechter* 15496 (holotype B).

Stem 8–13 cm long. *Leaves* 3.5 cm long, 1 cm wide, sheathing stem, dark green to yellow green becoming paler at margins and base. *Inflorescence* 9–12 cm long. *Flowers* opening at apex of inflorescence first, orange to yellow-green; lip trilobed, 1.5 mm long, 1.3 mm wide, lateral lobes erect, auriculate, midlobe triangular, broadest at apex with a small apical sinus, margins erose, base fleshy. (See fig. 12).

DISTRIBUTION: Ambae, Anatom, Efate, Erromango, Pentecost and Tanna. Also in the Solomon Islands, New Caledonia, Fiji, Samoa, Tahiti and Tonga.
HABITAT: Rain forest, 290–500 m.
COLLECTIONS: *Cribb* 53 (K); *Cribb & A. Morrison* 1784 (K); *Morrison* in RBG Kew 20, 48, 49, 50 & 153 (K); *Raynal* 15994 (P); *Schmid* 3624 (P); *Wheatley* 79 & 104 (K, PVNH).

Oberonia kaniensis Schltr. from New Guinea is similar and may prove to be conspecific.

2. O. heliophila *(Reichb. f.) Reichb. f.* in Drake, Ill. Fl. Ins. Mar. Pacif., Fasc. 7: 305 (1892).
Malaxis heliophila Reichb. f. in Otia Bot. Hamb.: 56 (1878). Type: Samoa, *Whitmee* s.n. (holotype K!).
Oberonia betchei Schltr. in Bull. Herb. Boiss., ser 2, 6: 303 (1906). Type: Samoa, *Betche* 38 (isotype MEL).

Leaves grouped at the base, 10–35 cm long, 1–1.5 cm wide. *Inflorescence* 20–30 cm long. *Flowers* white to pale brown with a yellow, pale or dark green lip; lip entire, obovate, 1.5 mm long, 1.2 mm wide, apex obtuse. (See fig. 12).

DISTRIBUTION: Efate. Also in the Solomon Islands, Fiji and Samoa.
HABITAT: Coastal to montane rain forest.
COLLECTIONS: *Cribb & A. Morrison* 1770 (K); *Morrison* in RBG Kew 47 (K).

3. O. imbricata *(Blume) Lindley*, Gen. Sp. Orch. Pl.: 17 (1930).
Malaxis imbricata Blume, Bijdr. Fl. Ned. Ind.: 395 (1825). Type: Java, *Blume* s.n. (holotype L).

Stems erect or pendent, up to 60 cm long. *Leaves* 1–2 cm long, c. 0.5 cm wide, mid-green becoming paler at margins and apex, appressed to stem. *Rhachis* 7–15 cm long. *Flowers* russet-ochre to orange or white; lip 1 mm long, 0.8 mm wide, lateral lobes erect, auriculate, midlobe oblong. (See fig. 12).

Fig. 12. *Oberonia heliophila*. **A**, habit × ⅔; **B**, flower × 14. *O. titania*. **C**, habit × ⅔; **D**, flower × 14. *O. equitans*. **E**, habit × ⅔; **F**, flower × 14. *O. imbricata*. **G**, habit × ⅔; **H**, flower × 14. **A** drawn from *Morrison* in RBG Kew 47; **B** from *Cribb & A. Morrison* 1770 (Kew spirit no. 42263); **C** from *Cheesman* 90; **D** from *Green* 2451; **E** from *Morrison* in RBG Kew 48; **F** from *Cribb & A. Morrison* 1784 (Kew spirit no. 42264); **G** from *Hallé* 16379; **H** from *Robinson* K133 (Kew spirit no. 44797). All drawn by Sue Wickison.

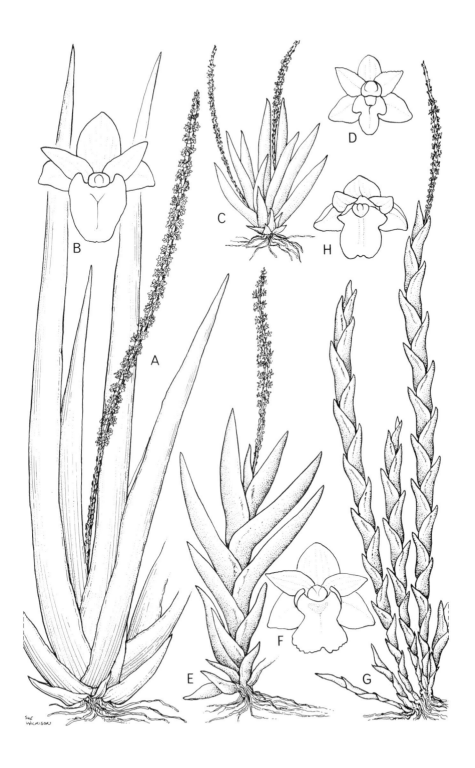

SUE
WICKISON

DISTRIBUTION: Anatom, Efate, Espiritu Santo, Pentecost and Tanna. Also in the Malay archipelago, Bougainville and the Solomon Islands.

HABITAT: Primary forest, 310–1200 m.

COLLECTIONS: *Cabalion* 1990 (K, P); *Cheesman* 15 (K); *Cribb & Wheatley* 8 & 26 (K, PVNH); *Green* in RSNH 1188 & 1346A (K); *Hoock* s.n. (P); *Morat* 5925 (P); *Morrison* in RBG Kew 53 & 56 (K); *Raynal* in RSNH 16370 (K, P); *Robinson* K133 (K); *Schmid* s.n. (P); *Wheatley* 182 (K, PVNH).

4. O. titania *Lindley*, Fol. Orch. Oberonia 1: 8 (1859). Type: Norfolk Island, *F. Bauer* s.n. (holotype W).

Titania miniata Endl., Prod. Fl. Norf.: 31 (1833). Type: Norfolk Island, *F. Bauer* s.n. (holotype W).

Oberonia palmicola F. Muell., Fragm. Phyt. Aust. 2: 24 (1860). Type: Australia, *H. Beckler* s.n. (holotype MEL).

Malaxis palmicola (F. Muell.) F. Muell., Fragm. Phyt. Aust. 7: 30 (1869).

Oberonia neocaledonica Schltr. in Engler, Bot. Jahrb. 39: 61 (1906). Type: New Caledonia, *Schlechter* 14766 (isotypes P!, Z, K!).

Leaves 3.5–6.0 cm long, c. 0.5 cm wide, yellow-green, grouped at the base. *Inflorescence* up to 11 cm long. *Flowers* red or green; lip trilobed, 0.8 mm long, 0.7 mm wide, midlobe ovate, lateral lobes narrowly trapeziform. (See fig. 12).

DISTRIBUTION: Ambae, Erromango and Espiritu Santo. Also in New Caledonia, Lord Howe Island and Norfolk Island.

HABITAT: Bush, 160–1200 m.

COLLECTIONS: *Cabalion* 2796 (P, PVNH); *Cheesman* 90 (K); *Wheatley* 86 (K, PVNH).

35. **ERIA** Lindley

Epiphytic. Pseudobulbs short or long, slender or thick and fleshy. *Leaves* distichous or near apex only, basal part always covered by sheaths. *Inflorescences* lateral or apparently terminal, 1– many-flowered. *Flowers* usually rather small, with a saccate to spur-like mentum; pollinia 8.

A genus of about 350 species from tropical Asia to Australia and the Pacific Islands. A single species in Vanuatu.

E. rostriflora *Reichb. f.* in Seem., Fl. Vit.: 301 (1868). Type: Fiji, *Seemann* 615 (holotype W; isotypes K!, P!).

Eria vieillardii Reichb. f. in Linnaea 41: 86 (1877). Type: New Caledonia, *Vieillard* 1335 (holotype P!).

Eria drakeana Kraenzl. in Viertelj. Nat. Ges. Zur. 74: 92 (1929). Type: New Caledonia, *Daeniker* 929 (holotype Z).

Eria kajewskii Ames in Journ. Arn. Arb. 13: 135 (1932); **synon. nov.** Type: Vanuatu, Anatom, *Kajewski* 820 (isotype K!).

Roots fibrous, elongated, branching, finely pubescent. *Pseudobulbs* stemlike, 6–24 cm long, clavate, concealed by brownish tubular, closely appressed sheaths, the upper portion bearing 5–6 crowded leaves. *Leaves* oblong, 6–18 cm long, 1.2–3 cm wide, unequally bilobed at apex, dark green. *Inflorescences* arising from the upper part of the pseudobulb, interspersed among the leaves, c. 9 cm long. *Flowers* c. 15 in a loose raceme, cream with a pale yellow lip; pedicels c. 1 cm long; dorsal sepal oblong, acute, c. 11 mm long; lateral sepals narrowly triangular, acute, 11 mm long, forming a short obtuse mentum with the column-foot; petals oblong-lanceolate, 9–10 mm long; lip ovate, acute, 5 mm long, 2.5 mm wide, margin with several blunt teeth on either side above the middle, 3-veined with the outer veins branching, with 2 basal calli; column, including foot, 5 mm long. (See plate 7c).

DISTRIBUTION: Ambae, Anatom, Efate, Erromango, Espiritu Santo, Pentecost and Tanna. Also in the Mariana Islands (Guam), the Solomon Islands, the Santa Cruz Islands, Fiji and Tahiti.

HABITAT: Rain forest of *Metrosideros* etc., 230–1760 m.

COLLECTIONS: *Bregulla* 13 (PVNH); *Cabalion* 1530 (PVNH) & 2768 (K, P, PVNH); *Cheesman* A83 (BM); *Cribb & A. Morrison* 1776 (K); *Cribb & Wheatley* 47 (K, PVNH); *Green* s.n. (K); *Kajewski* 820 (K); *MacKee* 32701 (P); *Morat* 5917 (P); *Raynal* 15990 & 16602 (P); *Suprin* 278 (P); *Chew Wee-Lek* in RSNH 244 (K, P); *Wheatley* 87 & 126 (K, PVNH).

36. **TRICHOTOSIA** Blume

Epiphytic. Plants covered in red–brown hairs. *Stems* long or short, leafy throughout, except at base. *Inflorescences* lateral; bracts at right angles to the rhachis, large and concave. *Flowers* not opening widely, with a prominent mentum; pollinia 8.

A genus of about 50 species from tropical Asia to the Pacific Islands. A single species in Vanuatu.

T. vulpina (*Reichb. f.*) *Kraenzl.* in Engl., Pflanzenreich Orch.-Mon.-Dendrob.: 141 (1911).

Eria vulpina Reichb. f. in Bonpl. 3: 222 (1855). Type: Philippines, *Cuming* 2071 (holotype K!).

Eria vanikorensis Ames in Journ. Arn. Arb. 13: 135 (1932); **synon. nov.** Type: Santa Cruz Islands, Vanikoro, *Kajewski* 512 (holotype AMES).

Plant covered in red-brown hairs. *Stems* up to 1 m tall, sheathed by leaf bases. *Leaves* distichous, oblong-lanceolate up to 12.5 cm long, 1.7–2.5 cm wide. *Inflorescences* 5 or more, lateral on the upper part of the stem opposite the leaves, c. 3.5 cm long; bracts c. 1 cm long, rigid. *Flowers* covered in red hairs on outer surface, pale green, probably self-pollinating; dorsal sepal lanceolate, c. 1 cm long; lateral sepals triangular-lanceolate, acute, forming a prominent obtuse mentum; petals linear-oblong, 8 mm long; lip c. 1 cm long, 3 mm wide, constricted 2 mm below the apex forming a terminal reniform or transversely elliptic lobe, disc with a short fleshy median callus. (See plate 3a).

DISTRIBUTION: Erromango and Espiritu Santo. Also in the Philippines and the Santa Cruz Islands.

HABITAT: Rain forest and montane ridge-top forest, 150–700 m.

COLLECTIONS: *Bregulla* 13 (PVNH); *Cabalion* 1530 & 2768 (PVNH); *Cribb & Wheatley* 63 (K, PVNH); *Green* in RSNH 1307 (K, P) & 1351 (PVNH); *Raynal* in RSNH 16221 (K, P, PVNH).

37. **MEDIOCALCAR** J.J. Smith

Small *epiphytic* or rarely terrestrial herbs. *Rhizomes* creeping, when young concealed by brownish bracts. *Pseudobulbs* conical or obliquely conical, close together or well spaced along the rhizome, 1- or 2-leaved, enveloped at base by several bracts. *Leaves* midgreen, slightly succulent. *Inflorescences* 1–2, 1-flowered, terminal; peduncles and pedicels elongate after fertilization. *Flowers* orange or red with green, white or yellow tips; peduncles often red; sepals ovate-lanceolate, connate in basal half; lateral sepals saccate at base; petals linear-lanceolate, free; lip entire, ovate, acute, clawed, saccate; pollinia 8.

A genus of 40–50 species from New Guinea and the adjacent islands. Two species in Vanuatu, often found growing together.

Leaves oblong to oblong-lanceolate, 1–2.5 cm wide; pseudobulbs with 1 leaf
... ... **2. M. paradoxum**
Leaves linear, 0.6–1.2 cm wide; pseudobulbs with 1 or 2 leaves
... ... **1. M. alpinum**

1. M. alpinum *J.J. Smith.* in Bull. Jard. Bot. Buitenz. ser 2, 13: 62 (1914). Types: New Guinea, *Kock* 68 & 143 (syntypes BO).
Mediocalcar bifolium var. *validum* J.J. Smith in Nova Guinea 12: 30 (1913). Types: as for *M. alpinum.*

Pseudobulbs 0.5–1.5 cm long, 1- or 2-leaved. *Leaves* linear-lanceolate, 7–9 cm long, 0.6–1.2 cm wide. *Peduncle* 1–2 cm long. *Flowers* 7–9 mm long; red to orange with yellow tips with a yellow lip; lip ovate, acute, 7 mm long; column 6 mm long. (See fig. 13).

DISTRIBUTION: Ambae, Erromango and Espiritu Santo. Also in New Guinea and the Solomon Islands.
HABITAT: Montane forest, 500–1200 m.
COLLECTIONS: *Bernardi* 13346 (P); *Cribb & Wheatley* 51 & 131 (K, PVNH); *McKee* in RSNH 24160 (K); *Raynal* in RSNH 16389 (P); *Wheatley* 39 (K).

2. M. paradoxum (*Kraenzl.*) *Schltr.* in Fedde, Rep. Sp. Nov. 9: 96 (1910).
Eria paradoxa Kraenzl. in Engler, Bot. Jahrb. 25: 606 (1898). Type: Samoa, *Reinecke* 300 (holotype B).
Mediocalcar vanikorense Ames in Journ. Arn. Arb. 13: 136 (1932); **synon. nov.** Type: Santa Cruz Islands, Vanikoro, *Kajewski* 641 (holotype AMES).

Pseudobulbs 1–1.5 cm long, 1-leafed. *Leaf* variable, oblong or oblong-lanceolate, 3.7–9 cm long, 1–2.5 cm wide, dark green. *Peduncles* 2–3 cm long. *Flower* 8–10 mm long; red or orange, with yellow tips; lip 9 mm long, ovate, acuminate or acute; column 6 mm long. (See fig. 13, plate 3b).

Fig. 13. *Mediocalcar paradoxum.* **A**, habit × ⅔; **B**, column × 4; **C**, flower × 4; **D**, petal × 4; **E**, dorsal and lateral sepal × 4; **F**, anther cap × 6; **G**, lip × 4. *M. alpinum.* **H**, petal × 4; **J**, anther cap × 6; **K**, lip × 4; **L**, anther cap × 6; **M**, flower × 4; **N**, column × 4; **O**, dorsal and lateral sepal × 4; **P**, habit × ⅔. **A** drawn from *Wickison* 22; **B–G** from *Mitchell* 9 (Kew spirit no. 50825); **H–L** & **N–O** from *Hunt* 2034 (Kew spirit no. 28351); **M** from *Robinson* K197 (Kew spirit no. 49172); **P** from *Hunt* 2030. All drawn by Sue Wickison.

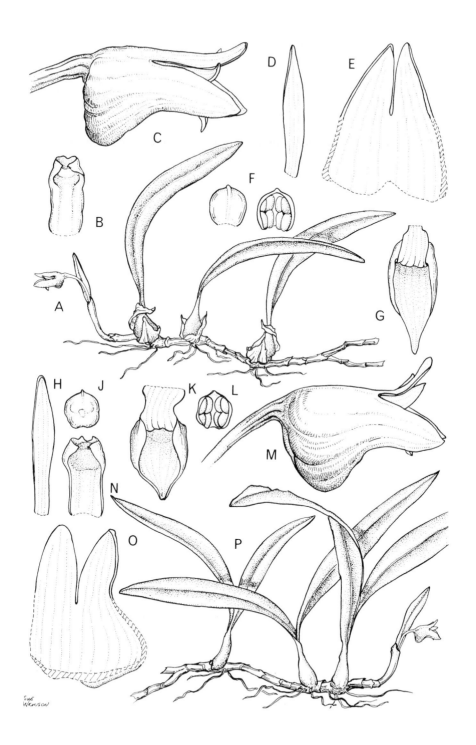

SUE WRKISON

DISTRIBUTION: Anatom, Efate, Erromango, Espiritu Santo and Tanna. Also in the Solomon Islands, the Santa Cruz Islands (Vanikoro), Fiji and Samoa.
HABITAT: Montane forest, 600–1550 m.
COLLECTIONS: *Bernardi* 13129 (G, P, K); *Cabalion* 1256 & 2784 (PVNH), 2782 (K); *Cheesman* 71 (K); *Cribb & Wheatley* 13, 39, 60, 86 & 92 (K, PVNH); *Green* in RSNH 1153 (K); *Morat* 6037 (P); *Morrison* in RBG Kew 109 (K); *Raynal* in RSNH 16155 (P); *Suprin* 279 (PVNH).

Some authorities recognise a third species, *Mediocalcar robustum* Schltr., in Vanuatu, which has a more robust growth and broader, larger leaves. However, there does not appear to be any clear distinction between the specimens we have seen and those of *M. paradoxum*.

38. **CERATOSTYLIS** Blume

Epiphytic. Stems slender, with thin brown sheaths at the base. *Leaf* solitary, apical, fleshy, narrow or almost terete. *Inflorescence* terminal, short of few–many small flowers, usually with a few open together. *Flowers* small, with sepals and petals about equal; lateral sepals forming a mentum enclosing the base of the lip; lip narrow at base, the blade thickened, entire; column short, divided at the apex into 2 erect arms which carry the stigmas on their inner surfaces; pollinia 8.
Young plants may have additional short green leaves at the base of the stem in place of some of the usual brown sheaths.
A genus of about 60 species from tropical Asia to the Pacific Islands. A single species in Vanuatu.

C. subulata *Blume*, Bijdr. Fl. Ned. Ind.: 306 (1825). Type: Java, *Blume* s.n. (holotype L).
Ceratostylis cepula Reichb. f. in Bonpl. 5: 53 (1857). Type: Java, *Zollinger* 3185 (holotype W).
Ceratostylis malaccensis Hook. f., Fl. Brit. Ind.: 823 (1890). Type: Malay peninsula, *Scortechini* (holotype K!).

Stems closely tufted, 7–15 cm long, dark green, variable in thickness from wiry to slender, 0.3–1.5 mm thick. *Leaf* solitary, almost terete, broadly grooved on one side, 3–5 cm long, 1–1.5 mm wide. *Flower* often solitary, dull red-purple to pale purple, with a white-pink mentum and a yellow lip; sepals ovate, c. 3 mm long, recurved at tips; lateral sepals forming a finely hairy mentum; petals lanceolate, acute, c. 2 mm long; lip narrow at the base, ovate, blunt, 3.5 mm long, 1.5 mm wide at fleshy apex, spur globose, c. 1 mm long, 1 mm wide. (See plate 4b).

DISTRIBUTION: Ambae, Anatom, Efate, Erromango, Espiritu Santo, Maewo and Pentecost. Also in the Malay peninsula and archipelago, Bougainville and the Solomon Islands.
HABITAT: Rain forest, 250–700 m.
COLLECTIONS: *Bernardi* 13190 (K, P); *Bourdy* 607 (P, PVNH); *Cabalion* 2865 (P, PVNH); *Cribb & Wheatley* 72 (K, PVNH); *Kajewski* 234 (K, P); *MacGillivray* 931 (K); *McKee* in RSNH 24157 (K, P); *Morrison* in RBG Kew 43, 102, 103, 106 & 107 (K); *Raynal* in RSNH 15991 (P) & 16144 (K, P, PVNH); *Schmid* 3978 (P); *Wheatley* 72 & 190 (K, PVNH).

Ceratostylis micrantha Schltr. from New Caledonia is similar and may prove to be conspecific.

39. **EPIBLASTUS** Schlechter

Epiphytic, large, sturdy, herbs. *Pseudobulbs* 1-leaved, enclosed by increasingly larger bracts. *Leaves* linear, flat, but at the base folded along the midrib. *Inflorescences* many, terminal, enveloped by base of leaf. *Flowers* waxy, reddish; dorsal sepal elliptic; lateral sepals obliquely triangular, forming a blunt open mentum; petals free; lip clawed, ovate, erect, trilobed, apex recurved, lateral lobes curving upwards, midlobe with a fleshy lateral ridge; column short and stout; pollinia 8.

A genus of about 15 species in New Guinea and the adjacent islands. A new genus record for Vanuatu, a single species being recorded.

E. sciadanthus (*F. Muell.*) *Schltr.* in K. Schum. & Laut., Nachtr. Fl. Deutsch. Sudsee: 137 (1905).
Dendrobium sciadanthum F. Muell. in Wing's Southern Sc. Rec. 2: 95 (1882). Type: Samoa, *Betche* s.n. (holotype MEL).

Stem not branched, consisting of 5–6 superposed pseudobulbs, yellow-green. *Leaves* linear-oblong, acute, c. 35 cm long, 3.2 cm wide, mid- to yellow-green. *Peduncle* 7–10.5 cm long. *Flowers* in clusters of up to 8, red at base fading out to pink at tips; pedicels c. 17 mm long; sepals 7–8 mm long; petals elliptic, 7–8 mm long; lip trilobed, 7.8 mm long, 4–5 mm wide when flattened, midlobe ovate, fleshy with a transverse fleshy ridge, lateral lobes erect, very broadly rounded, with the margins slightly incurved; column 2.8–3 mm long.

DISTRIBUTION: Ambae, Espiritu Santo. Also in the Solomon Islands, Fiji and Samoa.
HABITAT: Ridge-top forest, 150–800 m.
COLLECTIONS: *Cabalion* 2762 (NOU, PVNH), 2763 (P), & 2881 (P, PVNH); *Cribb & Wheatley* 56 (K, PVNH); *McKee* 32067 (P); *Wheatley* 74 (K, PVNH).

Eriks (1988), in an unpublished manuscript revision, considers the Vanuatu specimens to be a distinct species which has provisionally been named *Epiblastus brevipes*. However, from the drawings and description we have seen this does not appear to be distinct from *E. sciadanthus*.

40. **AGROSTOPHYLLUM** Blume

Epiphytic. *Stems* clustered, erect, of many internodes, bilaterally flattened, lower part with sheaths only, the leaves and flowers from the apical parts. *Leaves* distichous, in 2 ranks, with overlapping sheaths, twisted at the base to lie in one plane. *Inflorescence* terminal, with flowers in a globose apical head or in clusters on an elongate inflorescence, surrounded by small bracts. *Flowers* small, probably self-pollinating; sepals and petals similar, the petals narrower; mentum contains the saccate base of the entire lip; pollinia 8.

A genus of 40–50 species from tropical Asia and the Malay archipelago to the Pacific Islands. Five species are recorded from Vanuatu, four being new records:

Agrostophyllum costatum, A. graminifolium, A. leucocephalum and *A. torricellense*. The species are arranged in 3 sections according to Schlechter (1912).

1. Flowers in clusters on an elongate inflorescence ... ... **4. A. torricellense**
   Flowers in a globose head 2
2. Stems branching; leaves at right angles to the stem ... ...**5. A. costatum**
   Stems not branching; leaves at less than 90 degrees to the stem 3
3. Leaves strap-like, oblong, 10–14 cm long, 1.3–2 cm wide; stem thick and robust, 1–1.8 cm wide ... ... **3. A. majus**
   Leaves linear-lanceolate, 2.5–18 cm long, 2.5–9 mm wide; stem slender, 0.5–1 cm broad 4
4. Leaves long and grass-like, 12–18 cm long, 6–9 mm wide
   ... ...**1. A. graminifolium**
   Leaves 5–8.5 cm long, 2.5–4 mm wide   ... ...   **2. A. leucocephalum**

section AGROSTOPHYLLUM

1. A. graminifolium Schltr., in Fedde, Rep. Sp. Nov., Beih. 1: 266 (1912). Type: New Guinea, *Schlechter* 18594 (holotype B).

Stems 30–55 cm long, slender. *Leaves* linear, acute, 12–18 cm long, 0.6–0.9 cm wide, dark green; leaf bases 3–6 cm long. *Inflorescence* a globose head. *Flowers* white; sepals ovate, acute, c. 3 mm long; petals elliptic, obtuse, c. 3 mm long; lip obscurely trilobed in apical half, ovate, obtuse, 3 mm long, 1.5 mm wide, with 2 fleshy lateral calli; column c. 2 mm long.

DISTRIBUTION: Ambae, Banks Islands and Espiritu Santo. Also in Bougainville, New Guinea and the Solomon Islands.
HABITAT: Montane forest, 600–1500 m.
COLLECTIONS: *Cribb & Wheatley* 90 (K, PVNH); *Veillon* 5539 (P, PVNH); *Wheatley* 42 (K, PVNH).

2. A. leucocephalum *Schltr.* in K. Schum. & Laut., Nachtr. Fl. Deutsch. Sudsee: 128 (1905). Type: New Guinea, *Schlechter* 13998 (holotype B).

Stems 35–45 cm long, slender. *Leaves* linear, 5–8.5 cm long, 2.5–4 mm wide; leaf bases up to 3 cm long. *Inflorescence* a globose head. *Flowers* with ovate, acute sepals, up to 5 mm long; petals elliptic, acute, up to 5 mm long; lip obscurely trilobed in apical half, ovate, acute, 3 mm long, 1.5 mm wide; column c. 2 mm long.

DISTRIBUTION: Espiritu Santo. Also in New Guinea and New Caledonia.
HABITAT: Rain forest, 1100 m.
COLLECTION: *Cabalion* 2777 (in fruit) (P, K).

We have not been able to examine flowers of this species from Vanuatu, but in habit the specimen cited above is identical to the New Guinea specimens.
This species corresponds with *Agrostophyllum sp.*, of Hallé (1977).

3. A. majus *Hook. f.*, Fl. Brit. Ind. 5: 824 (1890). Type: India, collector in *King* s.n. (holotype K!).

Stems up to 1 m long, thick and robust. *Leaves* strap-like, obtuse, 10–14 cm long, 1.3–2 cm wide, mid-green; leaf bases up to 8 cm long. *Inflorescence* a globose head. *Flowers* white to yellow; sepals ovate, acute, 2–2.5 mm long; petals narrow, 2 mm long; lip ovate in apical half, obtuse, 2.5–3 mm long, 1.5 mm wide; column 2 mm long.

DISTRIBUTION: Erromango, Espiritu Santo and Pentecost. Widely distributed from Asia to the Malay archipelago, New Guinea and the Solomon Islands.

HABITAT: Wide variety of habitats from dense shade to almost full sun, 100–520 m.

COLLECTIONS: *Bourdy* 822 (P); *Cabalion* 749 & 2690 (P); *Cheesman* 96 (K); *Veillon* 2524 & 5520 (P); *Wheatley* 141 (K, PVNH).

Agrostophyllum parviflorum J.J. Smith from New Guinea is similar and may be conspecific.

Hallé (1986, herb. sheet) has noted the presence of elaters in the fruits of this species.

section DOLICHODESME

4. A. torricellense *Schltr.* in Fedde, Rep. Sp. Nov., Beih. 1: 260 (1912). Type: New Guinea, *Schlechter* 20205 (holotype B).

Stems 30–35 cm long. *Leaves* linear, acute, 17–20 cm long, 7–10 mm wide, dark green; leaf bases 6–8 cm long. *Inflorescence* elongate, 5–6 cm long, flowers in clusters of 3–5. *Flowers* not seen.

DISTRIBUTION: Ambae. Also in New Guinea.

HABITAT: Montane *Weinmania-Myrtaceae* forest, 1170 m.

COLLECTION: *Wheatley* 41 (K, PVNH) (in fruit).

section APPENDICULOPSIS

5. A. costatum *J.J. Smith* in Bull. Dép. Agric. Indes Néerl. 19: 1 (1908). Type: New Guinea, *Lorentz* 205 (holotype BO).

Rhizome elongated. *Stem* erect or pendulous, densely leaved, branched, up to 60 cm long, sheathed in imbricate leaf bases which make a characteristic zig-zag pattern on the stem. *Leaves* distichous, at right angles to the stem, oblong with a rounded apex, 1.8–2.1 cm long, 6–8 mm wide, mid-green, thin-textured. *Inflorescence* a globose head. *Flowers* white with a hint of green, with a white lip and lateral lobes with a maroon stripe; dorsal sepal ovate, 3.5 mm long; lateral sepals obliquely triangular, 2.8 mm long; petals linear, 2.6 mm long; lip trilobed, midlobe ovate, saccate, 3.5 mm long, 3.5 mm wide, lateral lobes erect, triangular, 2 mm long, 2 mm wide; column c. 2.5 mm long.

DISTRIBUTION: Banks Islands (Gaua). Also in New Guinea, Bougainville, the Solomon Islands and the Palau Islands.

HABITAT: No information from Vanuatu. In the Solomon Islands it is found in lowland forest in fairly open habitats, for example epiphytic on trees overhanging a river.

COLLECTION: *Bourdy* 966 (P) (sterile).

The description of the flowers is from Solomon Island specimens.

41. **APPENDICULA** Blume

Epiphytic, lithophytic or terrestrial. *Stems* not swollen at base, closely tufted, simple or branched, sheathed in leaf bases. *Leaves* distichous. *Inflorescences* lateral or terminal. *Flowers* small, green to white, may be flushed with maroon; dorsal sepal free; lateral sepals forming a mentum; pollinia 6.

A genus of about 50 species from tropical Asia to the Pacific Islands. Three species are recorded from Vanuatu, *Appendicula bracteosa* and *A. polystachya* being new records.

1. Inflorescence terminal, raceme elongate   … … … … … … … … … … … … 2
   Inflorescence lateral, raceme reduced   … … … … … … … … … **3. A. reflexa**
2. Bracts ovate, 4–6.5 mm long, 1.2–3 mm wide; sepals 4–5 mm long
   … **1. A. bracteosa**
   Bracts linear ovate, 1–1.5 mm long, 0.5 mm wide; sepals 2.5–3.5 mm long
   … … … … … … … … … … … … … … … … … … … **2. A. polystachya**

1. A. bracteosa *Reichb. f.* in Seem. Fl. Vit.: 299 (1868). Type: Fiji, *Seemann* 592 (holotype K!)

Stems up to 40 cm long, concealed by tubular bases of leaves, light green. *Leaves* distichous, oblong-lanceolate, acute, 3–4.0 cm long, 0.8–1.2 cm wide, dark green. *Inflorescence* terminal, branched basally or not, 7–11 cm long; bracts ovate, reflexed, 4–6.5 mm long, 1.2–3 mm wide. *Flowers* many, pale yellow-green with a yellow lip with a reddish base and a reddish-purple column; dorsal sepal ovate, 4 mm long; lateral sepals obliquely triangular, 5 mm long; mentum blunt; petals oblong-lanceolate, 4 mm long; lip obscurely trilobed, 5.5 mm long, 3 mm wide, lateral lobes erect, semicircular, midlobe oblong, with a horse-shoe shaped callus basally.

DISTRIBUTION: Pentecost. Also in Bougainville, the Solomon Islands, Fiji and Samoa.
HABITAT: Rain forest, 480 m.
COLLECTION: *Wheatley* 117B (K).
USES: In Fiji the roots of this species have been used as a poison.

This specimen was found growing with *Appendicula polystachya* (Schltr.) Schltr.

2. A. polystachya (*Schltr.*) *Schltr.* in Fedde, Rep. Sp. Nov., Beih. 1: 356 (1912).
Podochilus polystachyus Schltr. in K. Schum. & Laut., Nachtr. Fl. Deutsch. Sudsee: 121 (1905). Type: New Ireland, *Schlechter* 14664 (holotype B).

Stems up to 1 m long, erect or pendulous, concealed by tubular bases of leaves. *Leaves* distichous, oblong-lanceolate, acute, 4–4.5 cm long, 1–1.2 cm wide, light green to dark green, thin. *Inflorescence* terminal, branched basally, c. 12 cm long; bracts linear-ovate, reflexed, 1–1.5 mm long, 0.5 mm wide. *Flowers* many, greenish–white, may be flushed with maroon; lip green with a white apex; dorsal sepal ovate, c. 2.5 mm long; lateral sepals ovate, c. 3.5 mm long; mentum blunt; petals ovate, c. 2 mm long; lip oblong, broadest at apex, c. 3 mm long, 1.8 mm wide, disc with 2 broad, fleshy calli converging at base. (See fig. 14).

DISTRIBUTION: Anatom, Efate, Espiritu Santo and Pentecost. Also in New Ireland, Bougainville and the Solomon Islands.
HABITAT: Gorge and ridge-top forest, 20–800 m.
COLLECTIONS: *Cabalion* 1010 (PVNH), 2484 (P); *Cribb & Wheatley* 23 (K, PVNH); *Hoock* s.n. (P); *MacKee* 34544 (P); *Morat* 5477 (P); *Wheatley* 117A (K, PVNH).

3. A. reflexa Blume, Bijdr. Fl. Ned. Ind. 1(7): 301 (1825). Type: Java, *Blume* s.n. (holotype L, isotype P!).
Appendicula viridiflora Teijsm. & Binnend. in Nat. Tijdssk. Ned. Ind. 24: 321 (1862). Type: Java, *Teijsmann & Binnendijk* s.n. (holotype BO).
Appendicula vieillardii Reichb. f. in Linnaea 41: 76 (1877). Type: New Caledonia, *Vieillard* 1290 (holotype P!)
Appendicula cordata Hook. f., Fl. Brit. Ind. 6: 83 (1890). Type: Malay peninsula, *Scortechini* s.n. (holotype K!).
Podochilus reflexus Schltr. in Mém. Herb. Boiss. 21: 31 (1900). Type: Java, *Teijsmann & Binnendijk* s.n. (holotype BO).
Podochilus vieillardii (Reichb. f.) Schltr. in Engler, Bot. Jahrb. 39: 62 (1906).
Appendicula robusta Ridley, Fl. Malay Pen. 4: 197 (1924). Type: Malay peninsula, *Burkill & Holttum* 8856 (holotype SING).
Appendicula vanikorensis Ames in Journ. Arn. Arb. 13: 138 (1932); **synon. nov.** Type: Santa Cruz Islands, Vanikoro, *Kajewski* 594 (holotype AMES).
Appendicula dalatensis Guillaumin in Bull. Mus. Nat. Hist. (Paris) 6: 562 (1961). Type: Vietnam, *Sigaldi* 42 (holotype P).

Stems up to 60 cm long, concealed by sheathing tubular bases of the leaves. *Leaves* distichous, 2-ranked, twisted at the base so that the blades lie in one plane, oblong-lanceolate, 3.5–5.5 cm long, 8–15 mm wide, c. 1 cm apart, bluntly bilobed at apex. *Inflorescences* lateral, opposite leaves, c. 1 cm long, 2–4-flowered. *Flowers* white maturing yellowish; dorsal sepal narrowly elliptical, obtuse, concave, 2.5 mm long; lateral sepals broadly triangular, concave, c. 3 mm long; mentum blunt; petals oblong-lanceolate, c. 2 mm long; lip simple, elliptic, obtuse, 3 mm long, 2 mm wide, lightly saccate at the broad base, spur with a large transversely elliptic callus in the center; column and foot c. 2 mm long. (See fig. 14, plate 4a).

DISTRIBUTION: Anatom, Banks Islands (Gaua and Vanua Lava), Erromango and Espiritu Santo. Widely distributed from the Malay peninsula and archipelago to New Guinea, New Ireland, Bougainville, the Solomon Islands, the Santa Cruz Islands, the Palau Islands, the Horn Islands, New Caledonia and Fiji.
VERNACULAR NAME: Orvum-nge-nompull.
HABITAT: Poor red soil and limestone hills amongst bracken and under bushes, 160–800 m.
COLLECTIONS: *Bourdy* 430 (P) & 966 (PVNH); *Cabalion* 1364 (P, PVNH) & 1938 (P); *Cribb & Wheatley* 58 (K, PVNH); *Kajewski* 326 & 915 (K, AMES); *Quaife* 6 (K); *MacGillvray* 933 (K); *Morat* 7457 (P); *Suprin* 367 (P); *Veillon* 5529 (P); *Wheatley* 370 (K, PVNH).

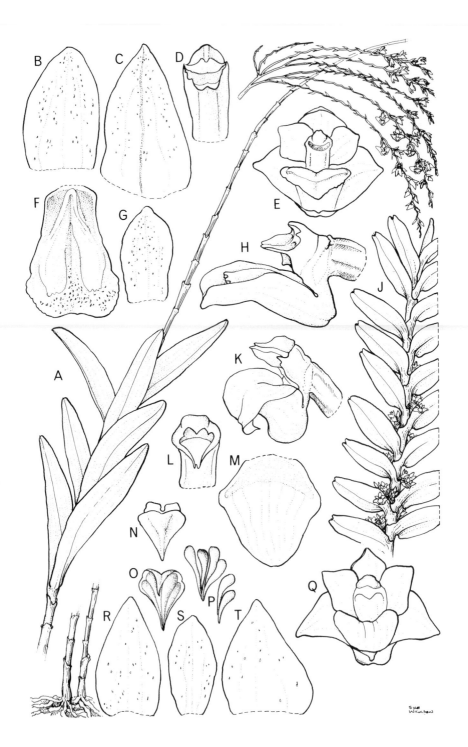

3

42. **OCTARRHENA** Thwaites

Epiphytic or rarely terrestrial. *Stems* elongate, leafy, branching from the root base. *Leaves* terete or laterally flattened. *Inflorescence* lateral, with small clustered flowers. *Flowers* with sepals and petals spreading; lip entire, about equal in length to the petals; column with no foot; pollinia 8.

A genus of about 20 species from tropical Asia to the Pacific Islands. A new genus record for Vanuatu, a single species being recorded.

O. angraecoides (*Schltr.*) *Schltr.* in Fedde, Rep. Sp. Nov., Beih. 1: 217 (1911).
Phreatia angraecoides Schltr. in K. Schum & Laut., Nachtr. Fl. Deutsch. Sudsee: 185 (1905). Type: New Guinea, *Schlechter* 14502 (holotype B).

Stem up to 20 cm tall, may appear zig-zag. *Leaves* fleshy, bilaterally flattened, 3–5 cm long, 1.5–3 mm wide, dark green. *Inflorescences* up to 5, 4–6 cm long, pale green, with many small dense flowers. *Flowers* c. 30, yellow; sepals ovate, c. 0.8 mm long; petals elliptic, c. 0.6 mm long; lip ovate, 0.6 mm long, 0.2 mm wide. (See plate 4d).

DISTRIBUTION: Erromango and Espiritu Santo. Also in New Guinea, Bougainville and the Solomon Islands.
HABITAT: Submontane ridge-top forest, 600–800 m.
COLLECTIONS: *Cribb & Wheatley* 34 (K, PVNH); *Green* in RSNH 1352 (K); *Robinson* K123 (K).

43. **PHREATIA** Lindley

This genus contains both pseudobulbous and non-pseudobulbous species, the former usually have 1–3 leaves, the non-pseudobulbous species may have as many as 12 leaves. *Epiphytic. Stem* absent, short or elongated. *Inflorescences* with many tiny flowers open together. *Flowers* often self-pollinating, white or green; lateral sepals form a distinct mentum; column-foot well developed; pollinia 8.

The tiny flowers of *Phreatia* makes their classification difficult, but vegetative characters can be used to key out most of the species.

A genus of about 150 species from S.E. Asia to Australia and the Pacific Islands. Six species in Vanuatu, *Phreatia caulescens* being a new record.

1. Plants pseudobulbous; leaves 1 or 2, from apex of pseudobulb; inflorescence from the base of pseudobulb … … … … … … … … … … … … … … … … … … 2
   Plants not pseudobulbous; leaves 4–12, distichous or arranged in a fan; inflorescences lateral … 3

Fig. 14. *Appendicula polystachya* **A**, habit × $\frac{2}{3}$; **B**, dorsal sepal × 10; **C**, lateral sepal × 10; **D**, column × 10; **E**, flower × 8; **F**, lip × 10; **G**, petal × 10; **H**, column and lip × 10. *A. reflexa.* **J**, habit × $\frac{2}{3}$; **K**, lip and column × 10; **L**, column × 10; **M**, lip × 10; **N**, anther cap top view × 14; **O**, anther cap ventral view × 14; **P**, pollinia × 14; **Q**, flower × 8; **R**, dorsal sepal × 10; **S**, petal × 10; **T**, lateral sepal × 10. **A–H** drawn from *Cribb & Wheatley* 23 (Kew spirit no. 53085); **J–T** from *Wickison* 44 (Kew spirit no. 50901). All drawn by Sue Wickison.

2. Leaves 12–24 cm long, 1.7–2.1 cm wide; lip ovate–oblong, acute, shortly apiculate, papillate at base ... ... **5. P. paleata**
Leaves (2.5) 7.5–11 cm long, 0.4–0.6 cm wide; lip spade-shaped, not papillate ... ... **2. P. hypsorhynchos**
3. Plant robust; leaves ribbon-shaped, 7–35 cm long, 1.5–2.5 cm wide ... ... **4. P. micrantha**
Plant delicate; leaves linear to linear-oblanceolate, 2.5–8.5 cm long, 0.1–0.6 cm wide 4
4. Stem elongate, up to 10 cm high ... ... **1. P. caulescens**
Stem less than 6 cm high 5
5. Leaves 3–7 cm long, 0.3–0.6 cm wide; lip shortly clawed, broadly ovate ... ... **6. P. stenostachya**
Leaves 2.5–8.5 cm long, 0.1–0.3 cm wide; lip square to oblong ... ... **3.P. matthewsii**

1. P. caulescens *Ames*, Orchid. 2: 200 (1908). Type: Philippines, *E.D. Merrill* 4587 (holotype K!).
Octarrhena caulescens (Ames) Ames, Orchid. 5: 192 (1915).

Stem leafy, lacking pseudobulbs, up to 10 cm long. *Leaves* distichous, linear, 7–8.5 cm long, 0.3–0.45 cm wide, bases up to 1.6 cm long. *Inflorescence* lateral, 9–17 cm long; peduncle 6.5–8 cm long; bracts linear, acuminate, 2–2.5 mm long. *Flowers* white; scented; dorsal sepal lanceolate, 2 mm long; lateral sepals ovate, 2.5 mm long; petals ovate–oblong, 2 mm long; lip subcircular, concave, 2.5 mm long, 2 mm wide; column 1 mm long.

DISTRIBUTION: Espiritu Santo. Also in the Philippines.
HABITAT: Summit forest on Mt. Tabwemasana, 1800 m.
COLLECTION: *Raynal* in RSNH 16333 (K, P) (sterile).

The description of the flowers is based on specimens from the Philippines.

2. P. hypsorhynchos *Schltr.* in Engler, Bot. Jahrb. 39: 77 (1906). Type: New Caledonia, *Schlechter* 15427 (isotypes K!, P!, Z).
Phreatia comptonii Rendle in Journ. Linn. Soc. 45: 249, t. 13, fig. 10–12 (1921).
Type: New Caledonia, *Compton* 580 (holotype BM!).

Sympodial, forming large clumps. *Pseudobulbs* proximate, ovoid, 4–7 mm high, 3–6 mm wide. *Leaves* 1 or 2, linear to oblanceolate, obtuse, erect, (2.5) 7–11 cm long, 0.4–0.6 cm wide. *Inflorescence* from base of pseudobulb, 10–22 cm long, much longer than the leaves; bracts linear-acuminate, 2–4 mm long. *Flowers* many in apical half, self-pollinating, white; pedicels wiry, 9–15 m long, ovary 3–5 mm long; sepals elliptic-ovate, obtuse, 1–1.5 mm long; mentum short, blunt and obscure; petals elliptic, obtuse, 1 mm long; lip spade-shaped, obtuse, 1.2 mm long, 1.5 mm wide.

DISTRIBUTION: Erromango and Espiritu Santo. Also in the Solomon Islands and New Caledonia.
HABITAT: Montane rain forest, 700–1800 m.
COLLECTIONS: *Bourdy* 190 (PVNH); *Cabalion* 2813 (K, P); *Cribb & Wheatley* 22 (K, PVNH); *Raynal* in RSNH 16381 (K, P).

3. P. matthewsii *Reichb. f.* in Otia Bot. Hamb.: 55 (1878). Type: Society Islands, *Matthews* s.n (holotype W).

Phreatia minutiflora sensu Kraenzl. non Lindley, in Engler, Bot. Jahrb. 25: 607 (1898).

Phreatia inversa Schltr. in K. Schum. & Laut., Nachtr. Fl. Deutsch. Sudsee 187 (1905). Type: New Ireland, *Schlechter* 14644 (holotype K!).

Phreatia reineckei Schltr. in Fedde, Rep. Sp. Nov. 9: 109 (1911). Type: Samoa, *Vaupel* 658 (holotype K!).

Phreatia neocaledonica Schltr. in Engler, Bot. Jahrb. 39: 78 (1906). Types: New Caledonia, *Schlechter* 14755 & 15228 (isosyntypes K!, P!, Z).

Monopodial. *Stems* short, clustered. *Pseudobulbs* absent. *Leaves* arranged in a fan, linear, obtuse, slightly fleshy, 2.5–8.5 cm long, 0.1–0.3 cm wide. *Inflorescence* lateral, equalling or shorter than leaves, many-flowered, 2.5–8 cm long; peduncle 1–3 cm long; bracts 1–1.5 mm long, reflexed. *Flowers* white or green, self-pollinating; sepals ovate, obtuse, 1–1.3 mm long; petals elliptic, subacute, 1 mm long; lip square or oblong, entire, c. 1 mm long, 0.5 mm wide.

DISTRIBUTION: Anatom, Banks Islands (Vanua Lava), Efate, Erromango, Espiritu Santo, Malekula and Pentecost. Also in New Ireland, Bougainville, the Solomon Islands, the Horn Islands, Fiji and Samoa.

HABITAT: Epiphyic on Kauri (*Agathis* spp.) etc., in dense rain forest, 310–500 m.

COLLECTIONS: *Cabalion* 2730 (K, P) & 3020 (P); *Cheesman* RBG Kew 5 (K); *Cribb & A. Morrison* 1777 & 1802 (K); *Cribb & Wheatley* 25 & 129 (K, PVNH); *Green* in RSNH 1127 (K, P); *Morrison* in RBG Kew 34, 35 & 36 (K); *Raynal* 15987 & 16268 (P); *Wheatley* 112, 254 & 365 (K, PVNH).

This species may be conspecific with *Phreatia myosurus* (G. Forst.) Ames, but that appears to differ in having broader leaves.

4. P. micrantha *(A. Rich.) Schltr.* in Fedde, Rep. Sp. Nov. 1: 919 (1913).

Oberonia micrantha A. Rich. in Sert. Astrol.: tab. 3 (1833). Type: Santa Cruz Islands, Vanikoro, *A. Richard* s.n. (holotype P!).

Phreatia macrophylla Schltr. in Engler, Bot. Jahrb. 39: 78 (1906). Type: New Caledonia, *Schlechter* 15465 (isotypes BM, K!, P!, W ,Z).

Phreatia sarcothece Schltr. in Fedde, Rep. Sp. Nov. 9: 438 (1911); **synon. nov.** Type: Vanuatu, Anatom, *Morrison* s.n. (holotype B).

Phreatia collina Schltr. in Fedde, Rep. Sp. Nov., Beih. 1: 919 (1913). Type: New Guinea, *Schlechter* 16438 (holotype B) non J.J. Smith (1911).

Phreatia robusta R. Rogers in Trans. Roy. Soc. Austr. 54: 39 (1930). Type: Australia, *A. Beck* s.n. (holotype AD).

Rhynchophreatia micrantha (A. Rich.) N. Hallé in Fl. Nouv. Caled. 8: 341 (1977).

Epiphytic, up to 40 cm tall. *Stem* very short, covered by persistent, imbricate leaf bases. *Leaves* alternate, 4–10, arranged in a fan, linear, blunt, 7–35 cm long, 1.5–2.5 cm wide, with an unequally bilobed apex. *Inflorescences* 1–several, from axils of the upper leaves, exceeding the leaves in length, erect or decurved; rhachis usually longer than the peduncle. *Flowers* numerous, minute, white; dorsal sepal ovate, cucullate, 1.5 mm long; lateral sepals obliquely ovate, 1.5 mm long; petals ovate, c. 1 mm long; lip subrhombic, 1.2 mm long, 1.2 mm wide, near the bottom there is a shallow depression containing viscid fluid; column c. 0.8 mm long, foot c. 0.7 mm long.

DISTRIBUTION: Ambae, Anatom, Banks Islands (Vanua Lava), Efate, Erromango, Espiritu Santo, Malekula, Pentecost and Tanna. Also in New Guinea, Bougainville, the Solomon Islands, the Santa Cruz Islands (Vanikoro), New Caledonia, Fiji, Samoa, the Horn Islands and Australia.

HABITAT: Rain forest, 200–1500 m.

COLLECTIONS: *Bourdy* 178 & 380 (K, P); *Bregulla* 34 (K); *Cheesman* H83/1930 (K); *Green* in RSNH 1349 (K); *Hallé* in RSNH 6412 (K, P); *Raynal* in RSNH 16175 (K, P); *Suprin* 370 (P); *Chew Wee-Lek* in RSNH 20 (K, P); *Wheatley* 96, 183 & 372 (K, PVNH).

This species was tranferred to the genus *Rhynchophreatia* by Hallé (1977), but we cannot see how it differs from the fan-leaved species of *Phreatia*.

5. P. paleata *(Reichb. f.) Reichb. f.* in Linnaea 41: 653 (1877).
Eria paleata Reichb. f. in Linnaea 41: 87 (1877). Type: New Caledonia, *Vieillard* 1331 (holotype P!).
Phreatia pholidotoides Kraenzl. in Not. Syst. 4: 140 (1928). Type: New Caledonia, *Cribs* 1217 (holotype P!).

Sympodial, clustered. *Pseudobulbs* obtuse, ovoid, 3–6 mm long, 3–5 mm wide. *Leaves* 1–2, oblanceolate, acute, 12–24 cm long, 1.7–2.1 cm wide, erect, petiolate. *Inflorescence* from base of pseudobulb, erect, 20–24 cm long, longer than the leaves, densely flowered in whorls, like a bottle-brush; bracts pale brown, spreading, 3.5–6 mm long. *Flowers* white; sepals ovate, acute, 3–4.5 mm long; mentum cylindrical, blunt, 1.5 mm long; petals ovate, acute, 2–2.5 mm long; lip ovate, acute or shortly apiculate, 2–2.8 mm long, 2–2.8 mm wide, papillate at base, spur shortly cylindrical, blunt, 0.8–1 mm long. (See plate 4e).

DISTRIBUTION: Banks Islands (Vanua Lava), Efate and Espiritu Santo. Also in New Caledonia and Norfolk Island.

HABITAT: Rain forest and ridge-top forest, 500–1200 m.

COLLECTIONS: *Cheesman* RBG Kew 8 (K); *Cribb & A. Morrison* 1815 (K); *Cribb & Wheatley* 54 (K, PVNH); *McKee* 32280 (P); *Raynal* in RSNH 16371 (K, P).

This species is similar to *Phreatia tahitiensis* Lindley, from Tahiti, and *P. saccifera* Schltr. from New Ireland.

6. P. stenostachya *(Reichb. f.) Kraenzl.*, in Engler, Pflanzenr. Orch.-Monand.-Thelasin.: 29 (1911).
Eria stenostachya Reichb. f. in Seem., Fl. Vit.: 301 (1868). Type: Fiji, *Seemann* 589 (holotype W, isotype P!).
Phreatia oubatchensis Schltr. in Engler, Bot. Jahrb. 39: 79 (1906); **synon. nov.** Type: New Caledonia, *Schlechter* 15393 (isotype K!, P!).

Monopodial. *Stems* clustered, covered by leaf bases, up to 6 cm long. *Leaves* in a fan, 4–8, linear to linear-oblanceolate, obtuse, 3–7 cm long, 0.3–0.6 cm wide. *Inflorescence* lateral, often longer than the leaves, up to 12 cm long; peduncle short, 1–4 cm long; bracts reflexed, 1–2 mm long. *Flowers* white, self-pollinating; sepals ovate, subacute, 2 mm long; mentum obtuse, rounded, 0.5 cm long; petals elliptic, subacute; lip shortly clawed, broadly ovate, obtuse, 1.5 mm long, 1.2 mm wide.

DISTRIBUTION: Ambae, Efate and Espiritu Santo. Also in the Solomon Islands, New Caledonia, Fiji and Samoa.

HABITAT: Rain forest and strand forest, sea level to 1500 m.
COLLECTIONS: *Cribb & A. Morrison* 1803 (K); *Cribb & Wheatley* 69 & 130 (K, PVNH); *Morrison in RBG Kew* 31 (K); *Raynal in RSNH* 16268 (K); *Wheatley* 84 (K, PVNH).

44. **AGLOSSORHYNCHA** Schlechter

Stems branched. *Leaves* distichous. *Inflorescence* terminal, short, 2-flowered. *Flowers* with a concave lip, with margins inrolled and embracing column; pollinia 4.
A genus of about 10 species in New Guinea and the adjacent islands. A single species in Vanuatu.

A. biflora *J.J. Smith* in Bull. Dép. Agric. Ind. Néerl. 39: 1 (1910). Type: New Guinea, *von Romer* 1290 (holotype BO).

Epiphytic or *lithophytic. Stems* sturdy, branched, pendent or spreading, yellow to olive-green, up to 1 m long, sheathed in leaf bases. *Leaves* many, distichous, oblong, 5–7 cm long, c. 0.8 cm wide, equally bilobed at the apex, mid-green on upper surface, light green on lower surface. *Flowers* 2, terminal, white to greenish, with a green lip; pedicels c. 6 mm long, finely pubescent; sepals and petals ovate-lanceolate, 13–15 mm long; lip concave, ovate, acute, 1.2 cm long, 0.8 cm wide, with inrolled margins embracing column.

DISTRIBUTION: Espiritu Santo. Also in New Guinea, Bougainville and Fiji.
HABITAT: Montane forest, 1500–1800 m.
COLLECTIONS: *Cabalion* 2845 (K, P); *Cribb & Wheatley* 78 (K, PVNH); *Hallé in RSNH* 16332 (PVNH); *Raynal in RSNH* 16332 (K, P, PVNH); *Veillon* 2456 (P); *Chew Wee-Lek in RSNH* 284 (PVNH).

45. **EARINA** Lindley

Epiphtyic or terrestrial herbs. *Leaves* distichous, coriaceous. *Inflorescence* terminal; rhachis erect, stiff, with flowers borne on short compressed, distichous branches. *Flowers* small, white; sepals and petals free; lip arched; pollinia 4.
A small genus of about 5 species in the Pacific Islands. A single species in Vanuatu.

E. valida *Reichb. f.* in Linnaea 41: 96 (1877). Type: New Caledonia, *Vieillard* 1298 (holotype P!).
Earina samoensium F. Muell. & Kraenzl. in Oest. Bot. Zeit. 44: 211 (1894). Type: Samoa, *Betche* 55 (holotype B).
Agrostophyllum drakeanum Kraenzl. in Journ. Bot. 17: 422 (1903). Type: New Caledonia, *Baudouin* 347 (holotype P!).
Earina brousmichei Kraenzl. in Not. Syst. 4: 136 (1928). Type: New Caledonia, *Brousmiche* 987 (holotype P!).

Epiphytic. Pseudobulb ovoid, less than 10 cm high, hidden by sheathing persistent leaf bases. *Leaves* distichous, arranged in a fan, lanceolate-linear,

30–40 cm long, 0.8–1.2 cm wide, unequally bilobed at the apex; leaf bases thickened and imbricate. *Inflorescence* terminal, c. 50 cm long, raceme c. 12 cm long, many-flowered; each branch, reduced, 4–10 mm long, distichous, 1–4-flowered. *Flowers* clustered, c. 40, white; dorsal sepal ovate, c. 4.5 mm long; lateral sepals ovate, c. 4.5 cm long; petals elliptic, 4.4 mm long; lip trilobed, constricted in the middle, 5 mm long, 3 mm wide, concave at base, lateral lobes rounded, erect, midlobe, ovate, deflexed at apex; column c. 4.5 mm long.

DISTRIBUTION: Anatom, Banks Islands (Vanua Lava), Erromango and Espiritu Santo. Also in New Caledonia, Fiji and Samoa.

HABITAT: Rain forest, 300–1070 m.

COLLECTIONS: *Bourdy* 408 (P) & 1061 (PVNH); *Cabalion* 2362 (P) & 2873 (K, P); *Cribb & Wheatley* 53 (K, PVNH); *Kajewski* 844 (K); *Wheatley* 369 (K, PVNH).

46. **GLOMERA** Blume

Epiphytic. Inflorescence terminal, globose, with flowers in a dense cluster, each subtended by a bract. *Flowers* white, lip with a rose apex; lateral sepals connate at base, forming a short conical-saccate mentum, enveloping spur; lip with a deeply saccate semicylindric spur; column short; pollinia 4.

A genus of about 25 species from the Malay archipelago to New Guinea and the Pacific Islands. Two species in Vanuatu, *Glomera papuana* being a new record.

Leaf sheaths smooth; leaves 9.5–15 cm long   ... ...**1. G. montana**
Leaf sheaths warty; leaves 1.5–4.0 cm long   ... ...**2. G. papuana**

1. G. montana *Reichb.f.* in Linnaea 41: 77 (1877). Type: Fiji, *Milne* s.n. (holotype K!).

G. rugulosa Schltr. in Fedde, Rep. Sp. Nov., Beih. 1: 287 (1912). Type: New Guinea, *Schlechter 18200* (holotype B).

Stems up to 1 m high, branching; leaf sheaths smooth. *Leaves* lanceolate, 9.5–15 cm long, 1–1.5 cm wide, dark green. *Flowers* 15–25; sepals and petals oblong-ovate, of equal length, c. 9 mm long; lip oblong, rounded at apex, spur globose c. 3 mm long, 3 mm wide; column c. 3 mm long. (See fig. 15, plate 4f).

DISTRIBUTION: Ambae, Anatom, Banks Islands (Vanua Lava), Espiritu Santo, Maewo and Pentecost. Also in New Guinea, Bougainville, Solomon Islands, Fiji and Samoa.

VERNACULAR NAME: Tupaida

HABITAT: In montane rain forest, 480–1300 m.

COLLECTIONS: *Cabalion* 1978, 2554 & 2897 (P); *Cheesman* A58 (BM); *Cribb & Wheatley* 12 (K, PVNH); *Suprin* 368 & 568 (P); *Wheatley* 42, 91, 124 & 368 (K).

Fig. 15. *Glomera montana*. **A**, habit × ⅔; **B**, flower × 3; **C**, dorsal sepal × 4; **D**, petal × 4; **E**, lateral sepal × 4; **F**, column × 6; **G**, anther cap × 8; **H**, pollinia × 8; **J**, lip × 6; **K**, lip and column × 6. *G. papuana*. **L**, habit × ⅔; **M**, flower × 3; **N**, lip and column × 6; **O** , column × 8; **P**, pollinia × 8; **Q**, anther cap × 8; **R**, dorsal sepal × 4; **S**, petal × 4; **T**, lateral sepal × 4; **U**, lip × 6. **A–K** drawn from *Cribb & Wheatley* 12 (Kew spirit no. 53172); **L–U** from *Hunt* 2961 (Kew spirit no. 28926). All drawn by Sue Wickison.

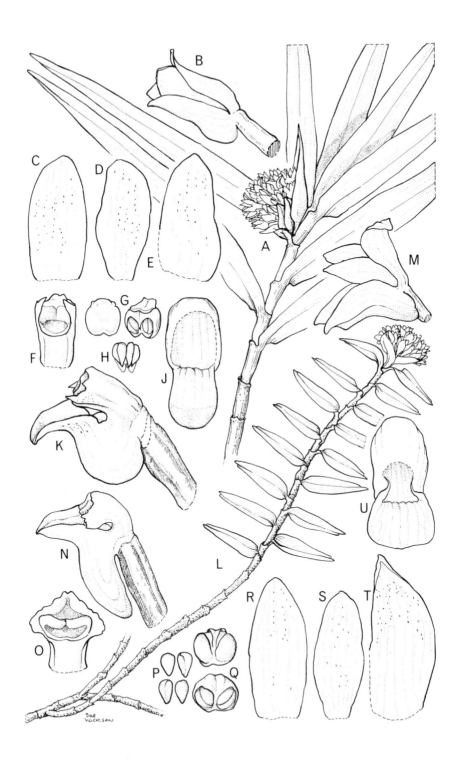

This species is similar to *Glomera samoensis* Rolfe from Samoa.

2. G. papuana *Rolfe* in Kew Bull. 1899: 111 (1899). Type: New Guinea, *A. Giulianetti* s.n. (holotype K!).

Stems 10–30 cm high, 1.5–2 mm wide, branching towards the base and covered by verrucose sheathing leaf bases. *Leaves* coriaceous, distichous, spreading, twisted at the base to lie in one plane, lanceolate, unequally bilobed at the subacute apex, 1.7–2.7 cm long, 3–8 mm wide; leaf bases sheathing. *Inflorescence* 1.5–2 cm in diameter; bracts elliptic, obtuse to acute, papery, up to 1.3 cm long. *Flowers* with pedicel and ovary 2 mm long; dorsal sepal oblong, obtuse, 9–10 mm long, 4 mm wide; lateral sepals obliquely oblong-ovate, acute, 10 mm long, 4.5 mm wide; petals narrowly elliptic, obtuse, 8 mm long, 3.5 mm wide; lip oblong to subquadrate, obtuse, 2–2.5 mm long, spur deeply saccate, semicylindrical, 2.5–3 mm long, 1.5–2 mm wide; column incurved, 2–2.5, with a short denticulate wing on each side at the apex. (See fig. 15).

DISTRIBUTION: Ambae and Espiritu Santo. Also found in New Guinea and the Solomon Islands.
HABITAT: Montane forest, 800–1550 m.
COLLECTIONS: *Cribb & Wheatley* 100 (K, PVNH); *McKee* in RSNH 24167 (PVNH); *Raynal* in RSNH 16335 (PVNH); *Veillon* in RSNH 4553 (K); *Chew Wee-Lek* in RSNH 243 (K); *Wheatley* 50 (K, PVNH).

These specimens differ slightly from the type of *Glomera papuana* in having blunter sepals and petals, a smaller obtuse lip and a longer spur which is dilated towards the apex. The Solomon Islands plants differ slightly from the Vanuatu specimens in having longer leaves up to 4 cm long.

47. **GLOSSORHYNCHA** Ridley

Epiphytic. Stem branching. *Leaves* slender, linear; upper border of the leaf sheath is ciliate. *Inflorescence* terminal, solitary, each flower subtended by a bract; lateral sepals free at base; pollinia 4.
A genus of about 50 species in New Guinea and the adjacent islands. A single species in Vanuatu.

G. macdonaldii *Schltr.* in Fedde, Rep. Sp. Nov. 3: 19 (1906). Type: Vanuatu, Anatom, *Macdonald* s.n. (holotype B).
Dendrobium sp. aff. *D. calcaratum* Lindley; Reichb. f. in Seem. Fl. Vit.: 303 (1868).
Dendrobium montis-movi Kraenzl. in Viertelj. Nat. Ges. Zurich 74: 87 (1929). Types: New Caledonia, *Daeniker* 544a, 544b & 544c (syntypes Z).
Glomera macdonaldii (Schltr.) Ames in Journ. Arn. Arb. 14: 111 (1933).
Dendrobium mouanum Guillaumin in Acta Horti Gotoburg. 18: 261, t. 26 (1950). Type: New Caledonia, *Skottsberg* 179 (holotype S).

Stem 12–50 cm tall, woody, branching, in dense spreading masses; leaf sheaths warty, with fimbriate margins. *Leaves* oblong-lanceolate, 2.5–3.0 cm long, c. 4 mm wide, with an unequally bilobed apex. *Flowers* white with yellowish tips to sepals and petals, sweetly fragrant; sepals and petals lanceolate, c. 12 mm long; lip apex triangular, spur cylindrical, c. 5 mm long, 2 mm wide; column c. 2.5 mm long. (See plate 4c).

DISTRIBUTION: Ambae, Ambrym, Anatom, Epi, Erromango, Espiritu Santo and Tanna. Also in Bougainville, the Solomon Islands, New Caledonia and Fiji.

HABITAT: Epiphytic on *Metrosideros* etc. in montane *Metrosideros-Weinmannia* forest, 700–1800 m.

COLLECTIONS: *Bernardi* 13161 (K, P, NOU); *Bourdy* 228 (P, PVNH); *Cabalion* 719 (PVNH), 2148 (P, PVNH), 2815 (K, PVNH); *Cheesman* 70 (K); *Cribb & Wheatley* 37 & 89 (K, PVNH); *Green* in RSNH 1151 (K, P, PVNH); *Morat* 6479 (P); *Morrison* in RBG Kew 38 (K); *Raynal* in RSNH 16145 & 16325 (P, K); *de la Rüe* s.n. (P); *Schmid* 3981 (NOU, P); *Suprin* 302 (P, PVNH); *Veillon* 4005 (P); *Chew Wee-Lek* in RSNH 220 (K, P); *Wheatley* 38 (K, PVNH).

48. **CADETIA** Gaudichaud

Epiphytic. Leaf solitary, terminal. *Flower* solitary, apical on a slender pedicel; sepals rather broad; lateral sepals form a mentum; petals narrower; lip trilobed, lateral lobes rather small, erect, midlobe variable but usually decurved; column with several erect teeth at apex; pollinia 4.

A genus of about 50 species from tropical Asia to Australia and the Pacific Islands. A new genus record for Vanuatu, a single species being recorded.

C. quadrangularis *Cribb & B. Lewis* in Orchid Rev. in press (1989). Type: Vanuatu, Efate, cult. *Begaud* s.n. (holotype K!).

Stem erect, up to 6 cm tall, sharply quadrangular in cross-section, enclosed in basal half by tightly fitting sheaths. *Leaf* erect or arcuate, coriaceous, linear to oblanceolate, obtuse, 5–7.5 cm long, 3–9 mm wide. *Inflorescence* one-flowered, axillary, very short; bract lanceolate, acute, 5–6 mm long. *Flower* white or creamy white; pedicel and ovary 1.2–1.4 cm long; dorsal sepal narrowly oblong-elliptic, obtuse, 4–4.5 mm long, 2–2.5 mm wide; lateral sepals obliquely oblong, rounded or obtuse, 5–5.5 mm long, 3 mm wide, forming a 2.5 mm long, saccate mentum at the base; petals linear, obtuse, 4–4.5 mm long, 0.8 mm wide; lip very fleshy, clawed at base, trilobed in the middle, 4–4.5 mm long and wide, lateral lobes erect, semicircular, rounded in front, midlobe very much larger than the lateral lobes, transversely oblong, elliptic, obtuse, 2 mm long, 4.5 mm wide; column 1.2 mm long, with a 2–2.5 mm foot. (See fig. 16).

DISTRIBUTION: Efate.

HABITAT: Forest and coconut plantations, sea level to 100 m.

COLLECTIONS: *Allen* s.n. & 2 (K); *Begaud* s.n. (K); *Bregulla* 12 (K); *Hallé* in RSNH 6296F (P); *MacKee* in RSNH 31421 & 34655 (P).

This species is readily distinguished from the common *Cadetia hispida* (A. Rich.) Schltr. from the Solomon Islands and Santa Cruz Islands by its stems, which are quadrangular in cross-section, and its flowers which have a glabrous ovary and lip. It is perhaps most closely allied to the New Guinea species *C. lucida* Schltr., but it seem to have longer, more sharply quadrangular stems and a glabrous lip and column.

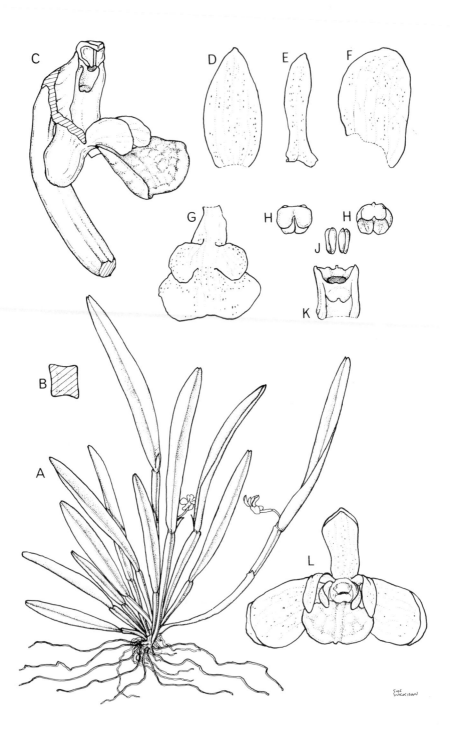

49. **DENDROBIUM** Swartz

Epiphytic, plants of extremely diverse form. *Pseudobulbs* present or absent. *Leaves* few–many, flat or terete, alternately arranged on stem, or crowded near apex of stem. *Inflorescences* lateral or terminal. *Flowers* solitary to numerous; sepals vary from short and broad to filiform; column with a pronounced column-foot forming a spur-like mentum with the bases of the lateral sepals; pollinia 2 or 4.

A large genus of over 1000 species from tropical and subtropical Asia to Australia and the Pacific Islands. *Dendrobium* is the largest genus in Vanuatu containing 28 species, 9 species being newly recorded there, namely, *D. spectabile, D. aegle, D. delicatum, D. masarangense, D. laevifolium, D. prostheciglossum, D. greenianum, D. kietaense* and *D. bilobum*. The species are arranged in 14 sections according to Schlechter (1912).

1. Flowers terminal or apparently terminal … … … … … … … … … … … 2
   Flowers lateral …13
2. Leaves terete and pendent … … … … … … … … … … … … … **1. D. seemannii**
   Leaves not terete and pendent … … … … … … … … … … … … … … … 3
3. Plants small, less than 3 cm tall with a creeping habit … … … … … … 4
   Plants large, more than 10 cm tall, erect or pendent … … … … … … … 5
4. Leaves ovate, 0.9–1.1 cm long, 0.5 cm wide; flowers red, purple, blue or
   yellowish-white … … … … … … … … … … … … … … … **13. D. delicatulum**
   Leaves linear, carinate, 3 cm long, 2 mm wide; flowers white, lip apex orange
   to golden-yellow … … … … … … … … … … … … … **14. D. masarangense**
5. Leaves crowded near apex of stem … … … … … … … … … … … … … 6
   Leaves distichous, spaced along stem … … … … … … … … … … … …10
6. Ovary and sepals hairy on outer surface … … … … … … … … … … … 7
   Ovary and sepals not hairy on outer surface … … … … … … … … … … 8
7. Pseudobulbs usually 2-leaved; leaves broad, 11.5–21 cm long, 6–11 cm wide;
   lateral lobes of lip tapering at apex, less than half as wide as midlobe
   … **5. D.polysema**
   Pseudobulbs usually 3-leaved; leaves 12–25 cm long, 4–8 cm wide; lateral
   lobes of lip subquadrate, wider across apex than base, as wide as midlobe
   … … … … … … … … … … … … … … … … … … … **3. D. macrophyllum**
8. Flowers opening widely; sepals and petals with undulate margins and
   somewhat twisted, c. 3.6 cm long; lip contorted, boldly purple-veined on all
   lobes … **6. D. spectabile**
   Flowers not opening widely; sepals and petals with smooth margins, not
   twisted, less than 2.2 cm long; lip not contorted … … … … … … … … 9
9. Sepals 7–10 cm long, yellow-green usually with red-brown blotching on
   sepals and lip … … … … … … … … … … … … … … … … … **2. D. macropus**
   Sepals 14–20 cm long, white; lip white with greenish venation
   … **4. D. mooreanum**

Fig. 16. *Cadetia quadrangularis.* **A,** habit × ⅔; **B,** cross section of stem × 3; **C,** side view of column, lip and spur × 8; **D,** dorsal sepal × 6; **E,** petal × 6; **F,** lateral sepal × 6; **G,** lip × 6; **H,** anther cap × 10; **J,** pollinia × 10; **K,** column × 12; **L,** flower × 4. Drawn from *Begaud* s.n. (Kew spirit no. 53267) by Sue Wickison.

10. Stems strongly constricted at base, swollen above, may appear zig-zag; leaves stiff, linear-lanceolate, bilaterally flattened, c. 0.5 cm wide ... ...**21. D. goldfinchii**
    Stems cane-like; leaves oblong, elliptic or ovate-elliptic, dorso-ventrally flattened, more than 0.8 cm wide 11
11. Sepals and petals yellow-green to lime-green and only slightly twisted; midlobe of lip straight; callus keels lilac, not raised at apex ... ...**20. D. macranthum**
    Sepal and petal colour not as above, greatly twisted; midlobe of lip recurved; callus keels raised at apex (1 or 3) 12
12. Flowers yellow-green to chocolate brown; petals twisted and contorted; lip with a callus of three ridges but only the central one raised at the apex ... ... **18. D. conathum**
    Flowers white to pale yellow; petals white, yellow, brown or violet, twisted; lip with a callus of five ridges the central three raised at the apex ... ... **19. D. gouldii**
13. Pseudobulbs large, 12–20 cm long, 3–4 cm wide, strongly bilaterally compressed ... ... **7. D. platygastrium**
    Pseudobulbs smaller than above, not bilaterally compressed 14
14. Inflorescence multiflowered 15
    Inflorescence 1- or 2-flowered 20
15. Flowers in dense globose heads, peduncle obscure **9. D. purpureum**
    Flowers on a short raceme; peduncle short but obvious 16
16. Flowers with lip-apex cucullate 17
    Flowers without a cucullate lip-apex 18
17. Flowers not opening widely, pink with white tips ... ... **11. D. aegle**
    Flowers open, brilliant red to red-orange ... ... **12. D. mohlianum**
18. Leaves ovate; lip trilobed, with a midlobe with 2 tails ... ... **16. D. prostheciglossum**
    Leaves oblong-lanceolate to lanceolate; lip entire 19
19. Flowers orange; leaves oblong-lanceolate, 9–16 cm long, 1.3–1.7 cm wide; stems robust ... ... **8. D. calcaratum**
    Flowers lilac-rose shading to white; leaves linear-lanceolate, 5–15 cm long, 0.5–1.2 cm wide; stems slender ... ... **10. D. rarum**
20. Plant small, less than 15 cm high; leaves from towards the apex of pseudobulb; lip entire ... ... **15. D. laevifolium**
    Plant more than 20 cm tall; leaves alternate on stem; lip trilobed 21
21. Leaves markedly decreasing in size towards the apex of stem; column with a motile appendage at base ... ... **27. D. insigne**
    Leaves not decreasing in size towards apex of stem; column not as above 22
22. Flowers solitary 23
    Flowers in pairs 24
23. Leaves oblong, 2–4 cm long, 0.5–0.9 cm wide; flowers resupinate, white to yellow to pale orange with tinges and stripes of darker orange; lateral lobes of lip oblong with a bifid apex ... ... **17. D. austrocaledonicum**
    Leaves lanceolate-oblong, 4–7 cm long, 0.4–0.5 cm wide; flowers non-resupinate, cream, pale yellow or yellowish-green; lip maroon or maroon and pale yellow; lateral lobes of lip semicircular ... ... **28. D. bilobum**
24. Leaves c. 3 mm wide, acute at apex ... ... **22. D. biflorum**
    Leaves more than 5 mm wide, bilobed at apex 25

section RHIZOBIUM

1. D. seemannii *L. O. Williams* in Bot. Mus. Leafl. Harv. Univ. 5: 123 (1938). Type: Fiji, *Seemann 579* (holotype K!).
Dendrobium crispatum sensu Reichb. f. non (G. Forst.) Sw. in Seem., Fl. Vit. 303 (1868).
Dendrobium calamiforme Rolfe in Kew Bull. 1921: 55 (1921) non Lodd. ex Lindley (1841).

Stems pendent, up to 2 m long. *Leaves* terete, 14–20 cm long, c. 2 mm wide, dark green; internodes 6–9 cm long. *Inflorescence* apical, from base of leaf, raceme 10–15 cm long. *Flowers* 7–15, cream-yellow, scented; sepals and petals lanceolate, c. 1.9 cm long; lateral sepals fused at base to form a mentum; lip oblanceolate, acuminate, recurved, c. 2.1 cm long, 0.5 cm wide, white spotted with purple on sides, calli 3, prominent at base, becoming flexuose on mid-lobe, smooth at apex; column c. 1 cm long, purple-green. (See plate 5a).

DISTRIBUTION: Efate, Erromango, Espiritu Santo and Malekula. Also in Fiji and the Society Islands.
HABITAT: Strand forest to ridge-top forest, usually in dry conditions and plenty of light, 10–700 m.
COLLECTIONS: *Bourdy* 138 (K, P, PVNH); *Bregulla* 4 (K); *Cabalion* 1641 (PVNH); *Cribb & A. Morrison* 1826 (K); *Cribb & Wheatley* 67 (K, PVNH); *Green* in RSNH 1111 (K, P, PVNH); *Hallé* in RSNH 6296B (P) & 6424 (K, P, PVNH); *McKee* 32278, 33723, 43482 & 43709 (P); *Sam* 150 (P, PVNH); *Seoule* 53 (P, PVNH); *Im Thurn* 325 (K); *Renz* 12540(G); *Robinson* K151 (K); *Walter* 404 (PVNH); *Chew Wee-Lek* in RSNH 276 (K, P, PVNH).

section DENDROCORYNE

2. D. macropus *(Endl.) Reichb. f. ex Lindley* in J. Linn. Soc. Lond., Bot. 3: 9 (1895).
Thelychiton macropus Endl., Prodr. Fl. Norf.: 33 (1833). Type: Norfolk Island, *F. Bauer* s.n. (holotype W).
Dendrobium gracilicaule F. Muell., Fragm. Phyt. Austr. 1: 179 (1859). Type: Australia: Moreton Bay, *W. Hill* s.n. (holotype MEL).
Dendrobium floribundum Reichb. f. in Gard. Chron. n. ser. 4: 772 (1875). Type: Vanuatu, cult. *Bull* 114 (holotype W, isotype K!) non *D. floribunden* D. Don (1825).

Dendrobium gracilicaule var. *howeanum* Maiden in Proc. Linn. Soc. N.S.W. 24: 382 (1889). Type: Lord Howe Island, cult. *Maiden* s.n. (holotype not found).

Dendrobium comptonii Rendle in J. Linn. Soc. Lond., Bot. 45: 247 (1920). Type: New Caledonia, *Compton* 1151 (holotype BM!).

Dendrobium drake-castilloi Kraenzl. in Not. Syst. 4: 135 (1928). Type: New Caledonia, *Vieillard* 3287 (holotype P!).

Dendrobium oscari A.D. Hawkes & A.H. Heller in Lloydia 20: 122 (1957). Type: as for *D. floribundum*.

Tropilis comptonii (Rendle) Rauschert, in Fedde, Rep. Sp. Nov. 94(7–8): 470 (1983) nom. illeg.

Stems clustered, erect, 20–60 cm long, c. 0.8 cm wide, slightly swollen at the base. *Leaves* 3–6, crowded near stem-apex, oblong, acute, unequally bilobed, 6–10 cm long, 1.5–3 cm wide. *Inflorescence* terminal; raceme 7–12 cm long; peduncle usually shorter than the rhachis. *Flowers* 5–30, yellowish green with dark purple-brown blotching on the outside of sepals, scented; pedicel 1–1.5 cm long; sepals and petals 7–10 mm long; dorsal sepal oblong; lateral sepals and petals falcate-oblong; lip trilobed, reflexed, 7–8 mm long, 5–6 mm wide, lateral lobes obliquely triangular, acute, mid-lobe reniform, callus 3-ridged, extends from base to mid-lobe; column c. 3 mm long. (See plate 5c).

DISTRIBUTION: Espiritu Santo. Also in New Caledonia, Fiji, Australia, Lord Howe Island, Norfolk Island, and Kermadec Island.

HABITAT: Lowland bush and rain forest.

COLLECTIONS: *Bregulla* 37 (K); *Bull* 114 (K); *Green* 2374 & 2455 (K).

section LATOURIA

3. D. macrophyllum *A. Rich.* in Sert. Astrol.: 22 (1834). Type: New Guinea, *A. Lesson* (holotype P!).

Dendrobium veitchianum Lindley in Reg. Bot. 33: t.25 (1847). Type: Java, *Lobb* s.n. (holotype K!).

Dendrobium ferox Hassk. in Retzia 1: 1 (1855). Type: Java, *Hasskarl* s.n. (holotype BO).

Dendrobium macrophyllum var. *veitchianum* (Lindley) Hook. f. in Curtis's Bot. Mag. 93: t.5649 (1867).

Dendrobium gordonii S. Moore ex Baker in Journ. Linn. Soc. 20: 372 (1883). Type: Fiji, *J. Horne* 942 (holotype K!).

Dendrobium brachythecum F. Muell. & Kraenzl. in Oest. Bot. Zeit. 44: 161 (1894). Type: New Guinea, *Anderson* s.n. (holotype MEL!).

Dendrobium ternatense J.J. Smith in Bull. Dép. Agr. Ind. Néerl. 22: 26 (1909). Type: Moluccas, Ternate, *J.J. Smith* s.n. (holotype BO!).

Dendrobium psyche Kraenzl., in Engler, Pflanzenr. Orch. Mon.- Dendr. 1: 246 (1910). Type: Vanuatu, *Braithwaite* s.n. (holotype B).

Dendrobium tomohonense Kraenzl. in Engler, Pflanzenr. Orch.- Mon.-Dendr. 1: 244 (1910). Type: Celebes, *Sarrasin* 655 & 799 (syntypes B).

Dendrobium musciferum Schltr. in Fedde, Rep. Sp. Nov., Beih. 1: 494 (1914). Types: New Guinea, *Schlechter* 16920 (syntype B); *Schlechter* 18958 (syntype B, isosyntype K!); *Schlechter* 19455 (syntype B).

Latourorchis macrophylla (A. Rich.) Brieger in Schlechter, Die Orchideen ed. 3, 1 (11/12): 727 (1981).
Latourorchis muscifera (Schltr.) Brieger, l.c.

Pseudobulbs clustered, clavate or subclavate, rarely fusiform, up to 50 cm tall, drying yellow, 4–6-noded below leaves. *Leaves* 3 (–7), suberect, coriaceous, oblong or oblong-elliptic, obtuse, 15–31 cm long, 3.3–9 cm wide. *Inflorescence* terminal, erect, up to 40 cm long, often densely many–flowered. *Flowers* rather variable in size; yellow or greenish with purple spots or stripes on the lateral lobes and mid-lobe of the lip; pedicel and ovary setosely hairy, 2.5–4 cm long; sepals setose on outer surface; dorsal sepal oblong-ovate or oblong-lanceolate, acute, 2.1–2.6 cm long; lateral sepals obliquely triangular, acute or acuminate, 2.3–2.6 cm long; mentum obliquely conical, 1 cm long; petals oblanceolate or oblong-oblanceolate, acute, 1.8–2.2 cm long, with somewhat undulate margin; lip trilobed, strongly recurved, 1–2 cm long, 1.6–2.7 cm wide, lateral lobes erect, subquadrate-flabellate, dilated above, truncate, mid-lobe transversely oblong, apiculate, conduplicate, callus obscurely 3-ridged, white; column 3 mm long, foot 1 cm long. (See fig. 17).

DISTRIBUTION: Ambrym, Anatom, Banks Islands (Vanua Lava), Efate, Espiritu Santo, Maewo and Pentecost. Widespread from the Malay archipelago to New Guinea and Bougainville, the Solomon Islands, Fiji and Samoa.
HABITAT: Rain forest, sea level to 1070 m.
COLLECTIONS: *Bernardi* 12976 (K, P) & 13191 (P); *Bourdy* 536 (PVNH); *Bregulla* 5 (K); *Cabalion* 2889 (P); *Cheesman* A21a (BM); *Cribb & A. Morrison* 1790 (K); *Green* 1050 (K); *Kajewski* 637 (AMES); *Raynal* in RSNH 16237 (K, P); *de la Rüe* s.n. (P); *Slade* s.n. (K); *Wheatley* 261 & 352 (K, PVNH).

4. D. mooreanum Lindley in Journ. Roy. Hort. Soc. 6: 272 (1851). Type: Vanuatu, Anatom, *C. Moore* s.n. (holotype K!).
Dendrobium petri Reichb. f. in Gard. Chron. n. ser. 7: 107 (1877). Type: Vanuatu, *P. Veitch* s.n. (holotype W).
Dendrobium fairfaxii Rolfe in Gard. Chron. ser. 3, 5: 798 (1889). Type: Vanuatu, *Fairfax* s.n. (holotype K!) non Muell. (1872).
Dendrobium quaifei Rolfe ex Ames in Journ. Arn. Arb. 14: 109 (1933). Type: Vanuatu, Espiritu Santo, *Quaife* s.n. (holotype K!).
Dendrobium priscillae A.D. Hawkes in Lloydia 20: 122 (1957). Type: as for *D. fairfaxii* Rolfe.

Pseudobulbs clustered, erect, subclavate, angulate, up to 25 cm tall, 3–4 noded, yellow. *Leaves* 2–4 at apex, coriaceous, spreading, ovate-lanceolate, acute, up to 8.5 cm long, 2.4 cm wide. *Inflorescences* terminal, 1–3, erect to arcuate, few-flowered, up to 18 cm long. *Flowers* white with greenish venation on lip; pedicel and ovary 1.9–2.3 cm long; dorsal sepal lanceolate, acute, 1.4–2.0 cm long; lateral sepals recurved, obliquely lanceolate, acuminate, 1.8–2.0 cm long; mentum obliquely conical, slightly incurved, 7.5 mm long; petals oblanceolate, acute to acuminate, 2.1–2.5 cm long; lip entire to obscurely trilobed, somewhat rhombic in outline, acute, 2.0–2.1 cm long, 0.8–1.2 cm wide, callus fleshy, half length of lip, 3-ridged; column 2 mm long, foot incurved, 7 mm long. (See plate 5b).

DISTRIBUTION: Ambae, Ambrym, Anatom, Banks Islands (Vanua Lava), Efate, Erromango, Espiritu Santo, Malekula and Tanna.

HABITAT: Lowland to montane forest, 300–1100 m.

COLLECTIONS: *Bernardi* 94, 95, 96, 97, 98 (K), 12867, 12919, 12991, 13067 (G), 12973 (G, P), 13350 (K), 13085(G, K, L, P); *Bourdy* 847 (PVNH); *Cabalion* 1970 (PVNH), 1543, 2759, 2872 & 2889 (P, PVNH); *Cribb* 38 (K); *Dixon* s.n. (K); *Fairfax* s.n. (K); *Green* in RSNH 1227 & 1163 (K); *MacKee* 41897 (P); *Morat* 7430 (PVNH); *Quaife* s.n.; *Raynal* in RSNH 16172 & 16194 (P, PVNH); *de la Rüe* s.n. (P); *Robinson* K129 (K); *Slade* s.n. (K); *Veitch* s.n.; *Wheatley* 37 345 (K, PVNH).

Dendrobium mooreanum closely resembles *D. ruginosum* Ames from the Solomon Islands differing in its relatively shorter, narrower petals and entire or very obscurely lobed rhombic lip with a triangular acute apex (Cribb, 1983).

5. D. polysema *Schltr.* in K. Schum. & Laut., Nachtr. Fl. Deutsch. Sudsee: 163 (1905). Type: New Guinea, *Schlechter* 14066 (holotype B).

Dendrobium macrophyllum var. *stenopterum* Reichb. f. in Gard. ser. 3,3: 393 (1888). Type: New Guinea, Hort. *Bull* (holotype W).

Dendrobium pulchrum Schltr. in Fedde, Rep. Sp. Nov., Beih. 1: 496 (1912). Type: New Guinea, *Schlechter* 19605 (holotype B).

Dendrobium polysema var. *pallidum* Chadim in Austr. Orch. Rev. 29(1): 32 (1964). Type: New Guinea, *Chadim* D7 (holotype K!).

Pseudobulbs clavate, 20–50 cm tall, drying yellow or orange, 3–5-noded below leaves. *Leaves* 2, spreading, coriaceous, elliptic or oblong-elliptic, obtuse, up to 22 cm long, 6–13 cm wide, shortly petiolate. *Inflorescence* terminal, erect to arcuate, few–many-flowered, up to 45 cm long. *Flowers* yellow or greenish, heavily spotted maroon, and with lateral lobes of lip striped maroon; pedicel and ovary densely setose, 3.6–5 cm long; sepals setose on outer surface; dorsal sepal ovate-lanceolate, acuminate, 2–2.4 cm long; lateral sepals obliquely falcate-lanceolate, acuminate, 2.3–3 cm long; mentum incurved-conical, 1.2 cm long; petals often reflexed, oblanceolate or obovate, apiculate, 1.8–2.6 cm long, with or without strongly undulate margins; lip trilobed, recurved, 1.8–2.3 cm long, 2–2.6 cm wide, lateral lobes smaller than mid-lobe, erect or incurved, tapering to a rounded apex, mid-lobe flat or with incurved margins, transversely oblong–ovate, subapiculate, callus 3-ridged, creamy white; column 3–3.5 mm long, foot incurved, 1.2 cm long. (See fig. 17, plate 5e).

DISTRIBUTION: Ambae, Efate, Erromango, Espiritu Santo and Pentecost. Also in New Guinea, Bougainville, the Solomon Islands and the Santa Cruz Islands.

HABITAT: Rain forest, 150–700 m.

COLLECTIONS: *Cabalion* 942 (PVNH); *Cribb & Wheatley* 66 (K, PVNH); *Green* in RSNH 1348 (K); *Raynal* in RSNH 16237 (K); *Robinson* K139 & K141 (K); *Wheatley* 92 & 250 (K, PVNH).

Fig. 17. *Dendrobium polysema.* **A,** habit × $\frac{2}{3}$; **B,** lateral sepal inside view × 3; **C,** lateral sepal outside view × 3; **D,** lip side view × 3; **E,** lip flattened × 3; **F,** column × 3; **G,** petal × 3; **H,** anther cap × 3; **J,** pollinia × 3; **K,** dorsal sepal × 3. *D. macrophyllum.* **L,** flower × 1; **M,** lip side view × 2; **N,** lip flattened × 2. A–K drawn from *Wickison* 38 (Kew spirit no. 50049); L–N from *Wickison* 38B (Kew spirit no. 28926). All drawn by Sue Wickison.

6. D. spectabile (*Blume*) *Miq.*, Fl. Ind. Bat. 3: 645 (1855).
Latourea spectabilis Blume, Rumphia 4: 41, t. 195 fig. 1 & t. 199 C (1850). Type: New Guinea, *Latour-Leschenault* s.n. (holotype L).
Dendrobium tigrinum Rolfe ex Hemsley in Ann. Bot. 5: 507 (1891). Type: Solomon Islands, San Cristobal, *Cumins* 187 (holotype K!).
Latourorchis spectabile (Blume) Brieger in Schlechter, Die Orchideen, ed. 3, 1 (11/12): 727 (1981).

Pseudobulbs clustered, cane-like, not noticeably clavate, up to c. 40 cm high, 5–8-noded below leaves, swollen at base. *Leaves* 4–6, suberect, coriaceous, elliptic, obtuse, up to 23 cm long, 4–8 cm wide, shortly petiolate at base. *Inflorescences* emerging from just below leaf-bases, erect, 20–40 cm long, few–many-flowered; peduncle terete. *Flowers* large, somewhat grotesque, yellow, commonly heavily mottled with maroon on sepals and petals, lined with maroon on lip, scented; pedicel and ovary 4–6 cm long; sepals with somewhat undulate margins; dorsal sepal recurved, lanceolate, acuminate, 3.6 cm long; lateral sepals recurved, lanceolate-falcate, acuminate, 3.6 cm long; mentum obliquely conical, 1 cm long; petals linear-lanceolate, acuminate, somewhat twisted, 4 cm long; lip trilobed, recurved, 4 cm long, 2.2 cm wide, lateral lobes erect, subquadrate-semicircular, rounded in front, mid-lobe much longer than lateral lobes, lanceolate, acuminate, undulate-twisted, callus 3-ridged, raised at base and apex, white; column short, 3 mm long, foot 8.5 mm long.

DISTRIBUTION: Espiritu Santo. Also in New Guinea, Bougainville and the Solomon Islands.
HABITAT: Lowland swampy forest, lower montane forest or on planted coconut or *Casuarina* trees, sea level to 1100 m.
COLLECTIONS: None in herbaria.

G. Cayrol (pers. comm.) collected plants of this species from east Espiritu Santo and these are currently in cultivation in Noumea, New Caledonia.

section PLATYCAULON

7. D. platygastrium *Reichb. f.* in Otia Bot. Hamb. 1: 55 (1878). Type: Fiji, *C. Wilkes* s.n. (holotype W).
Dendrobium platycaulon Rolfe in Kew Bull. 1892: 139 (1892). Type: Philippines, *Sander & Co* s.n. (holotype K!).

Young plants erect, mature plants pendent. *Pseudobulbs* clustered, basal part slender, upper leaf-bearing part enlarged, strongly bilaterally compressed, 12–30 cm long, 2–4 cm wide, acute at apex, dirty yellow to olive-brown; internodes 2–3 cm. *Leaves* alternate, from internodes at narrow edge of stem, up to 8, elliptic, 6.5–10 cm long, 2.5–4.5 cm wide, unequally bilobed, dark green, may be tinged mauve on underside. *Inflorescences* lateral, near apex of pseudobulb, 1–4, c. 1.5 cm long. *Flowers* 2–4, not opening widely, pale yellow to pink to purple; pedicels c. 2.5 cm long; sepals lanceolate, 1.5–1.8 cm long; petals ovate, c. 1.6 cm long; mentum cylindric, c. 1.3 cm long, at 45–90 degrees to pedicel; lip trilobed, 2–2.5 cm long, 0.8–1.4 cm wide, lateral lobes triangular, acute or truncate, white with pale pink lines, calli 3–5; column short, c. 4 mm long, foot c. 1.3 cm long, mauve. *Fruits* maroon with yellow seeds. (See plate 5f).

DISTRIBUTION: Efate and Erromango. Also in the Philippines, the Solomon Islands and Fiji.

HABITAT: Lowland and mountain forest in open localities, growing with moss and ferns, rare, sea level to 300 m.

COLLECTIONS: *Bregulla* 14 (K); *Cabalion* 1642 (PVNH); *Wheatley* 9 (K, PVNH).

The type material of *Dendrobium platycaulon* is very variable in size of flowers, angle of the spur to the pedicel and size and shape of the lip. This species is similar to *D. camptocentrum* Schltr., from New Caledonia.

section PEDILONUM

8. D. calcaratum *A. Rich.*, Sert. Astrol.: 18, t. 7 (1834). Type: Santa Cruz Islands, Vanikoro, *A. Richard* s.n. (holotype P-lost).
Dendrobium triviale Kraenzl. in Engler, Bot. Jahrb. 25: (1898). Type: Samoa, *Reinecke* 422 (holotype B).
Dendrobium separatum Ames in Journ. Arn. Arb. 13: 133 (1932). Type: Santa Cruz Islands, Vanikoro, *Kajewski* 503 (holotype AMES).

Stems clustered, pendent, ribbed, up to 1.5 m long, slightly swollen at the base, yellow-green turning maroon with age. *Leaves* alternate, on upper part of stem only, oblong-lanceolate, acute, 9–16 cm long, 1.3–1.7 cm wide, unequally bilobed at apex. *Inflorescences* lateral, pendent or ascending, from nodes near apex of old leafless stems, 2.0 cm long, mauve. *Flowers* 10–18, with basal flowers developing first, fleshy, bright orange; pedicel c. 1 cm long, green to mauve; dorsal sepal oblong, concave, c. 5 mm long; lateral sepals oblong, concave, 1.4 cm long; mentum cylindric, c. 9.5 mm long, appressed to ovary; petals elliptic-lanceolate with an erose upper margin, c. 5 mm long; lip with the saccate basal portion fused to the column-foot, free portion ovate, 1.3 cm long, 3 mm wide, with a transverse lamella dividing the basal and apical parts of the lip, oblong-elliptic at apex, with the margins inrolled and erose; column including foot 11 mm long. (See fig. 18, plate 6f).

DISTRIBUTION: Ambae, Banks Islands (Vanua Lava), Espiritu Santo and Pentecost. Also in New Britain, the Solomon Islands, the Santa Cruz Islands and Samoa.

HABITAT: Rain forest, sea level to 500 m.

COLLECTIONS: *Cabalion* 748 (NOU, P) & 2604 (P); *Cribb & Wheatley* 124 (K, PVNH); *Wheatley* 90, 249 & 321 (K, PVNH).

9. D. purpureum *Roxb.*, Fl. Ind. 3:484 (1832). Type: Moluccas, illustration in Rumphius Herb. Amb. 6: 110, t.50, f.1.
Angraecum purpureum Rumph., Herb. Amb. 6:110 t.50 f.1 (1750).
Dendrobium morrisonii Schltr. in Bull. Herb. Boiss., ser 2, 6: 456 (1906). Type: Vanuatu, Anatom, *Morrison* s.n. (holotype B).
Dendrobium sertatum Rolfe in Journ. Linn. Soc. 34: 174 (1909). Type: Fiji, *L.S. Gibbs* 610 (holotype K!).

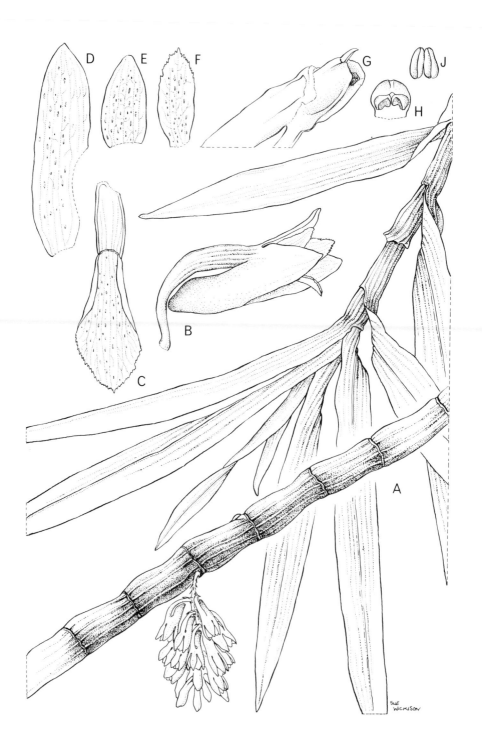

Stems clustered, pendent, ribbed, 4-angled, up to 50 cm long, slightly swollen at base; yellow-green turning purple with age. *Leaves* alternate, oblong-lanceolate, 7–14 cm long, 1.2–2.0 cm wide, unequally bilobed at apex, dark green. *Inflorescences* lateral, at right angles to the leafless stem, arising from nodes; peduncle 3–5 mm long; raceme very condensed. *Flowers* 10–15, purple, rose-purple or white; pedicel c. 1 cm long; dorsal sepal oblong-lanceolate, c. 8 mm long; petals lanceolate, c. 7mm long; lateral sepals oblong, c. 1.5 cm long; mentum cylindrical, appressed to ovary, c. 6 mm long; lip with the saccate basal portion fused with the column-foot, ovate, acute 1.0 cm long, 5 mm wide, apical margins erose, deflexed at apex; column including foot 1.0 cm long, white. (See fig. 19, plate 6d).

DISTRIBUTION: Ambae, Anatom, Efate, Espiritu Santo, Malekula and the Shepherd Islands (Tongoa). Also in the Malay archipelago, New Guinea, Bougainville, the Caroline Islands and Fiji.

HABITAT: Lowland to montane forest, 150–980 m.

COLLECTIONS: *Allen* 16 (K); *Bregulla* 3 (K); *Cribb* 55 (K); *Cribb & Wheatley* 73 (K, PVNH); *Hallé* in RSNH 6296C; *Im Thurn* 377 & 378 (K); *McKee* 32091 (P); *Morat* 6466 (P); *Morrison* in RBG Kew 44, 114, 115 & 116 (K); *Reeve* 543 (K); *Wheatley* 70 & 76 (K, PVNH).

This species has been reported to occur in Australia but it may have been confused with *Dendrobium smillieae* F. Muell. (Lavarack & Gray, 1985).

10. D. rarum *Schltr.* in Fedde, Rep. Sp. Nov., Beih. 1: 504 (1912). Types: New Guinea, *Schlechter* 18252 & 19512 (syntypes B).
Dendrobium calcaratum auct. non A. Rich. in Sert. Astrol.: 18 (1834).

Stems clustered, erect, ribbed, up to 1 m long, slightly swollen at base, dull purple, enclosed in persistent leaf sheaths. *Leaves* linear-lanceolate, acute to acuminate, unequally bilobed, 5–15 cm long, 0.5–1.2 cm wide, thin-textured, fresh green, becoming flushed dull purple above. *Inflorescences* lateral, pendulous, racemose, arising from nodes on the upper half of leafless stems, 2–4.5 cm long, olive-green or purple-green turning brick red. *Flowers* 10–15, lilac-rose shading to white distally, spreading; pedicel and ovary 1.3–1.7 cm long; sepals cucullate at apex; dorsal sepal ovate, acute, 0.6 cm long; lateral sepals obliquely triangular-ovate, acute, 1.3 cm long; mentum acute, slightly decurved at apex, 0.5–0.7 cm long; petals ovate, minutely papillose, transluscent white, 0.4–0.6 cm long; lip with a saccate basal portion fused with the column-foot, ovate, acute, minutely papillose, 0.8–0.9 cm long, 0.4–0.5 cm wide; column 0.2 cm long, white, column-foot 0.5–0.7 cm long, pale lilac-rose. (See fig. 19, plate 6e).

DISTRIBUTION: Anatom, Erromango, Espiritu Santo, Pentecost and Tanna. Also in New Guinea.

VERNACULAR NAME: Korwisyèl.

HABITAT: Primary and secondary forest, 40–800 m.

Fig. 18. *Dendrobium calcaratum*. **A**, habit × ⅔; **B**, flower × 3; **C**, lip × 4; **D**, lateral sepal × 4; **E**, dorsal sepal × 4; **F**, petal × 4; **G**, column × 6; **H**, anther cap × 6; **J**, pollinia × 6. All drawn from *Wickison* 141 (Kew spirit no. 52126) by Sue Wickison.

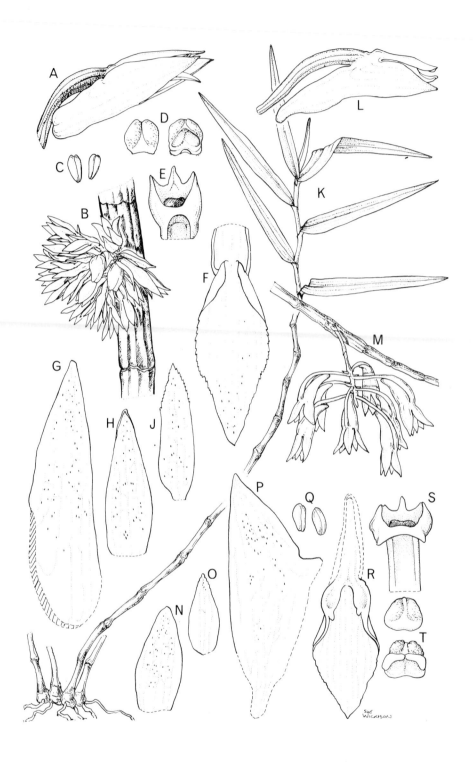

COLLECTIONS: *Bernardi* 13086 (G, P, K); *Cabalion* 777 (P), 1671 & 1736 (PVNH); *Cribb & A. Morrison* 1796 (K); *Cribb & Wheatley* 20 (K, PVNH); *Green* in RSNH 1263 (K, P); *Morrison* in RBG Kew 45, 113 & 118 (K); *Raynal* in RSNH 16225, 16603 (P), 16170 & 16226 (K, P); *Wheatley* 198 (K, PVNH).

A specimen cultivated by G.H. Slade may be referable to this species but it differs in having much paler flowers.

section CALYPTROCHILUS

11. D. aegle *Ridley* in Journ. Linn. Soc. 32: 260 (1896). Type: Java, *Ridley* s.n. (holotype BM).
Dendrobium aemulans Schltr. in K. Schum. & Laut., Nachtr. Fl. Deutsch. Sudsee: 176 (1905); **synon. nov.** Type: New Guinea, *Schlechter* 14329 (holotype B).
Dendrobium inopinatum J.J. Smith in Bull. Jard. Bot. Buitenz. ser. 3, 5: 88 (1922). Type: Sumatra, *Bunnemeijer* 1132 (holotype BO).
Pedilonum aegle (Ridley) S. Rauschert in Fedde, Rep. Sp. Nov. 94(7–8): 457 (1983).
Pedilonum inopinatum (J.J. Smith) S. Rauschert, l.c.

Stems clustered, pendent, ribbed, up to 40 cm long, not swollen at base, yellow-green turning purple with age. *Leaves* alternate, lanceolate, acute, 5–10 cm long, 0.8–1.5 cm wide. *Inflorescences* lateral, pendent, from nodes on leafless stems; peduncle 7 mm long. *Flowers* clustered, 5–10, not opening widely, pink with white tips; pedicels c. 2.3 cm long; dorsal sepal broadly ovate, 5 mm long; lateral sepals oblique-oblong, 1.2 cm long; petals obovate, 4 mm long; lip spathulate, 11 mm long, 2 mm long, margins incurved, with apical margin lacerate, spur cylindric, appressed to ovary, c. 1 cm long; column c. 3 mm long, foot c. 9 mm long. (See fig. 20).

DISTRIBUTION: Espiritu Santo. Also in Java, New Guinea, Bougainville and the Solomon Islands.
HABITAT: Ridge top forest, 750 m.
COLLECTIONS: *Cribb & Wheatley* 21 (K, PVNH); *Raynal* in RSNH 17005 (P-spirit).

12. D. mohlianum *Reichb. f.* in Bot. Zeit. 20: 214 (1862). Type: Fiji, *Seemann* 578 (holotype K!).
Dendrobium neo-ebudanum Schltr. in Bull. Herb. Boiss., ser 2, 6:456 (1906). Type: Vanuatu, *Morrison* s.n. (holotype B).
Dendrobium vitellinum Kraenzl. in Engl., Pflanzenr. Orch.- Mon.-Dendr. 124: 113, fig. 7 (1910); **synon. nov.** Type: Vanuatu, *MacDonald* s.n. (holotype B?).

Fig. 19. *Dendrobium purpureum*. **A**, flower × 3; **B**, inflorescence life-size; **C**, pollinia × 6; **D**, anther cap × 6; **E**, column × 6; **F**, lip × 4; **G**, lateral sepal × 4; **H**, dorsal sepal × 4; **J**, petal × 4. *D. rarum*. **K**, habit × ⅔; **L**, flower × 3; **M**, inflorescence life-size; **N**, dorsal sepal × 4; **O**, petal × 4; **P**, lateral sepal × 4; **Q**, pollinia × 6; **R**, lip × 4; **S**, column × 6; **T**, anther cap × 6. **A–J** drawn from *Cribb* S5 (Kew spirit no. 43828); **K–T** from *Cribb & Wheatley* 20 (Kew spirit no. 53170). All drawn by Sue Wickison.

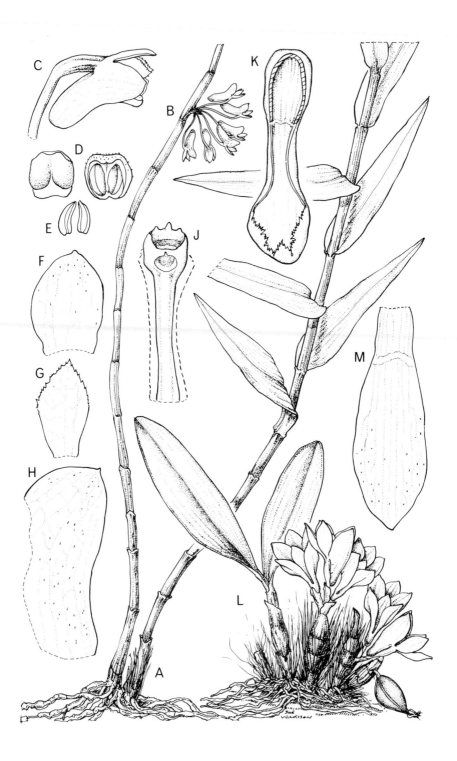

Stems spreading or pendent, clustered, ribbed, up to 50 cm long, not swollen at the base. *Leaves* distichous, sheathing the stem at the base, lanceolate, 6–13 cm long, 1–2.5 cm wide. *Inflorescence* lateral from leafless stem; peduncle 1–1.5 cm long. *Flowers* racemose, in clusters of 4–6, orange to bright red; pedicels 1.5–2.0 cm long; dorsal sepal and petals ovate, c. 9 mm long; lateral sepals triangular, c. 1.5 cm long, fused at the base to form a mentum; lip obovate, 16 mm long, 10 mm wide at apex, cucullate at apex, orange-red with or without purple lines; column c. 3 cm long, white, with foot c. 1.3 cm long. (See plate 6c).

DISTRIBUTION: Ambae, Ambrym, Anatom, Epi, Erromango, Espiritu Santo, Malekula and Tanna. Also in the Solomon Islands, Fiji and Samoa.
VERNACULAR NAME: Kworriyziwèr
HABITAT: Montane forest, 430–950 m.
COLLECTIONS: *Bernardi* 13015(G), 13137(G, P) & 13168 (K, G, P); *Bourdy* 862 (P); *Cabalion* 2147 (P); *Cheesman* 72 (K); *Cribb & Wheatley* 44 (K, PVNH); *Fuller* s.n. (K); *Green* in RSNH 1173 & 1224 (K, P); *Morrison* in RBG Kew 46, 80, 112, 117 (K) & s.n. (P); *Raynal* in RSNH 16187 (K, P), 16149 & 16354 (P); *de la Ruë* s.n. (P); *Wheatley* 30 (K, PVNH).

Kraenzlin confused *Dendrobium mohlianum* with *D. lawesii* F. Muell., from New Guinea and the Solomon Islands, although the two species are quite distinct.
Wheatley (pers. comm.) has observed that the flowers of this species become progressively redder at higher altitudes.

section OXYGLOSSUM

13. D. delicatulum Kraenzl. in Engler, Bot. Jahrb. 16: 17 (1893). Types: New Guinea, *Hellwig* 303 (holotype B-destroyed), *A. Millar* NGF 22862 (neotype LAE, isotype K!).
Dendrobium minutum Schltr. in Fedde, Rep. Sp. Nov., Beih. 1: 531 (1912). Type: New Guinea, *Schlechter* 18754 (holotype B).

Dwarf creeping plant forming dense mats. *Pseudobulbs* ovoid-globose, sometimes clavate, c. 5 mm long, 2.5 mm wide, yellow-green, bearing 2 leaves at the apex. *Leaves* ovate to elliptic, 0.9–1.1 cm long, c. 0.5 cm wide, glossy dark green. *Inflorescences* arising from both leafless and leafy stems (from between the leaves), 1–3-flowered. *Flowers* purple, blue, red or yellowish white, with lip apex usually bright orange to orange-red; not opening widely; pedicels c. 5 mm long; ovary indistinctly 5-ribbed; sepals and petals c. 5 mm long; sepals lanceolate; petals linear-lanceolate; mentum appressed to ovary, cylindric, obtuse, c. 5 mm long; lip oblong-linear, apex acute, reflexed, c. 7 mm long, 2 mm wide; column c. 3 mm long. (See fig. 21, plate 6b).

Fig. 20. *Dendrobium aegle*. **A**, habit × ⅔; **B**, inflorescence × ⅔; **C**, flower × 3; **D**, anther cap × 6; **E**, pollinia × 6; **F**, dorsal sepal × 4; **G**, column × 4; **H**, petal × 4; **J**, lateral sepal × 4; **K**, lip × 4. *D. laevifolium*. **L**, habit × ⅔; **M**, lip × 3. A & C–K drawn from *Cribb & Wheatley* 21 (Kew spirit no. 53171); **B** from Mitchell 6; **L–M** from *Cribb & Wheatley* 108 (Kew spirit no. 54287). All drawn by Sue Wickison.

DISTRIBUTION: Erromango and Espiritu Santo. Also in the Caroline Islands (Ponape Island), New Guinea, Bougainville, the Solomon Islands and Fiji.
HABITAT: Ridge-top montane forest in deep shade, 1400 m.
COLLECTIONS: *Cribb & Wheatley* 109 (K, PVNH); *Raynal* in RSNH 16601 (P).

These collections agree well with the typical subspecies (Reeve & Wood, 1981).

14. D. masarangense *Schltr.* in Fedde, Rep. Sp. Nov. 10: 78 (1911). Types: Celebes, *Schlechter* 20473 (holotype B); New Britain, *Ridsdale & Latik* NGF 38046 (neotype K!).
Dendrobium pumilio Schltr. in Fedde, Rep. Sp. Nov., Beih. 1: 471 (1912). Type: New Guinea, *Schlechter* 20267 (holotype B).

Tufted plants. *Pseudobulbs* clustered, ovoid, c. 1.0 cm long, 2.5 mm wide, light green. *Leaves* 2, apical, carinate, c. 3 cm long, 2 mm wide. *Flowers* apical, 1 or 2, white with a green base; pedicel c. 1 cm long; ovary indistinctly 5-ribbed; sepals and petals lanceolate, acuminate, 7–8 mm long; mentum narrowly conical, appressed to ovary, c. 5 mm long; lip oblong, acuminate, 11 mm long, 2 mm wide, orange to golden yellow at apex; column c. 3 mm long, green; anther cap pale green. (See fig. 21).

DISTRIBUTION: Ambae, Banks Islands (Vanua Lava), Erromango and Pentecost. Also in New Guinea, the Solomon Islands and New Caledonia.
HABITAT: Rainforest, 300–1350 m.
COLLECTIONS: *Bourdy* 125 (P); *Cabalion* 2991 (P, PVNH); *Raynal* 16601 (P); *Wheatley* 113 & 366 (K, PVNH).

section CUTHBERTSONII

15. D. laevifolium *Stapf* in Bot. Mag. 149: t. 9011 (1924). Type: New Guinea: Rossell Island, *J.C. Frost* s.n. (holotype K!).
Dendrobium occultum Ames in Journ. Arn. Arb. 14: 108 (1933). Type: Santa Cruz Islands, Vanikoro, *Kajewski* 604 (holotype AMES).

Pseudobulbs clustered, ovate-oblong, some elongated up to 4.5 cm high, 1 cm wide, covered in fibres of decayed cataphylls. *Leaves* 2–3, from apex of pseudobulb, oblong-lanceolate, 4–10 cm long, 1–3 cm wide, bright green. *Inflorescences* lateral, from shorter pseudobulbs; peduncle c. 1.5 cm long. *Flowers* usually in pairs (sometimes 1 or 3), pink or white, often with a yellow margin to the lip; sepals and petals c. 1.3 cm long; sepals ovate; lateral sepals c. 8 mm long; mentum obtuse; petals obovate; lip naviculate, c. 2.8 cm long, 0.9 cm wide when flattened, with yellow margins; column c. 4 mm high, minutely toothed; anther cap whitish. (See fig. 20, plate 6a).

Fig. 21. *Dendrobium delicatulum*. **A**, habit × 3; **B**, flower × 4; **C**, dorsal sepal × 6; **D**, petal × 6; **E**, lateral sepal × 6; **F**, lip × 6; **G**, column × 8; **H**, pollinia × 8; **J**, anther cap × 8. *D. masarangense*. **K**, habit × 3; **L**, flower × 4; **M**, anther cap × 8; **N**, lip × 6; **O** anther cap × 8; **P**, pollinia × 8; **Q**, dorsal sepal × 6; **R**, petal × 6; **S**, lateral sepal × 6; **T**, column × 8. **A–J** drawn from *Wickison* 23 (Kew spirit no. 50873); **K–T** from *Hunt* 2936 (Kew spirit no. 28596). All drawn by Sue Wickison.

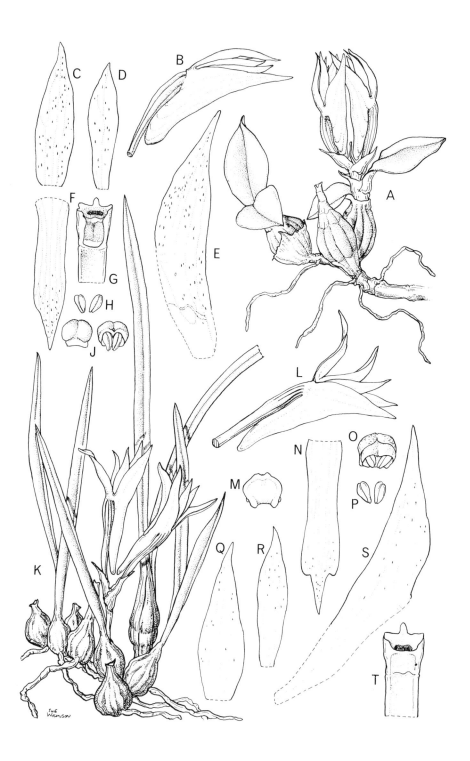

DISTRIBUTION: Ambae, Efate and Espiritu Santo. Also in Louisades Archipelago, the Solomon Islands and the Santa Cruz Islands.

HABITAT: Steep slopes in montane forest in deep shade, on small trunks near the ground, c. 1300–1400 m.

COLLECTIONS: *Bourdy* 126 (NOU); *Cabalion* 2728 (P); *Cribb & Wheatley* 108 (K, PVNH); *Wheatley* 43 (K, PVNH).

This species is similar to *Dendrobium prasinum* Lindley from Fiji.

section TRACHYRHIZUM

16. D. prostheciglossum *Schltr.* in Fedde, Rep. Sp. Nov., Beih. 1: 555 (1912). Type: New Guinea, *Schlechter* 18174 (holotype B, isotype K!).
Trachyrhizum prostheciglossum (Schltr.) F.G. Brieger in Schlechter, Orchideen ed. 3, 1(11–12): 687 (1981).

Epiphytic or occasionally *terrestrial*. *Stems* clustered, erect, ribbed, 20–90 cm tall; leaf bases sheathing (in dried specimens they may appear darker giving the stem a banded appearance); internodes 1–3.5 cm. *Leaves* alternate, ovate, 4–7 cm long, 1.5–2.2 cm wide, unequally bilobed at apex. *Inflorescences* lateral, from base of leaf sheath, c. 10 cm long. *Flowers* 4–9, yellow-green tinged on the outside with purple or brown; pedicels c. 3 cm long; dorsal sepal broadly ovate, c. 1 cm long; lateral sepals obliquely triangular, c. 1 cm; mentum obtuse; petals linear, c. 1 cm long; lip trilobed, 1.5 cm long, 1.6 cm wide, mid-lobe 2-tailed, white, lateral lobes triangular, callus at base horse-shoe-shaped; column c. 2 mm long, foot c. 6 mm long, upturned at apex.

DISTRIBUTION: Erromango. Also in New Guinea and New Britain.

HABITAT: Rain forest. No information from Vanuatu; in New Guinea it grows in montane forest, 1100–1200 m.

COLLECTION: *Bernardi* 13210 (G, K) (in fruit).

We have been unable to examine flowering material of this species and therefore the description of the flowers is from Schlechter (1912).

section DISTICHOPHYLLUM

17. D. austrocaledonicum *Schltr.* in Fedde, Rep. Sp. Nov. 3: 80 (1906). Type: New Caledonia, *Le Rat* 187 (holotype B).
Dendrobium inaequale Finet in Bull. Soc. Bot. Fr. 50: 375 (1903) non Rolfe (1901). Type: New Caledonia, *Balansa* 761 (holotype P!).
Dendrobium cerinum Schltr. in Engler, Bot. Jahrb. 39: 72 (1906) non Reichb. f. (1879). Type: as for *D. inaequale*.
Dendrobium critae-rubrae Guillaumin in Bull. Soc. Bot. Fr. 103: 281 (1956). Type: Vanuatu, Anatom, *L.E. Cheesman* s.n. (holotype BM!).
Dendrobium garayanum Hawkes & Heller in Lloydia 20: 19 (1957). Type: as for *D. inaequale*.

Fig. 22. *Dendrobium austrocaledonicum.* **A**, habit × ⅔; **B**, flower × 3; **C**, dorsal sepal × 3; **D**, lateral sepal × 3; **E**, petal × 3; **F**, column and mentum × 3; **G**, pollinia × 6; **H**, anther cap × 6; **J**, lip × 3; **K**, column × 4. A drawn from *Wickison* 12; **B–K** from *Sprunger* 129 (Kew spirit no. 51940). All drawn by Sue Wickison.

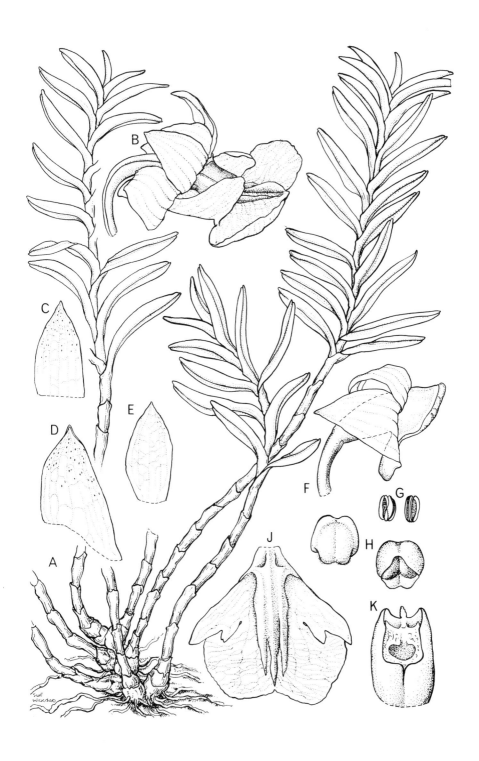

113

Dendrobium mendoncanum Hawkes in Orquidea 25: 19 (1963). Type as for *D. inaequale.*

Stems clustered, ribbed, 25–40 cm long, slightly swollen at base. *Leaves* distichous, sheathing stem at base, oblong, 2–4 cm long, 0.5–0.9 cm wide, light yellow-green, unequally bilobed at apex. *Flowers* lateral, solitary, probably self-pollinating; young flowers white, maturing yellow to pale orange with tinges and stripes of darker orange; pedicel and ovary c. 17 cm long; sepals and petals reflexed; dorsal sepal ovate, c. 8 mm long; lateral sepals ovate, c. 1. 3 mm long, fused at base to form a mentum; petals ovate, with an erose upper margin, c. 8 mm long; lip trilobed, c. 1.3 cm long, 1.2 cm wide; lateral lobes erect, oblong, unequally bifid at apex; mid-lobe reniform with a 3-ridged fleshy callus; column c. 4 mm long. (See fig. 22).

DISTRIBUTION: Anatom and Erromango. Also in Bougainville, the Solomon Islands and New Caledonia.
HABITAT: Rainforest, 10–400 m.
COLLECTIONS: *Bernardi* 12979 (G, P); *Cabalion* 1368 (PVNH); *Cheesman* s.n. (BM); *Hoock* s.n. (P).

section SPATULATA

18. D. conanthum *Schltr.* in Fedde, Rep. Sp. Nov., Beih. 1: 550 (1912). Type: New Guinea, *Schlechter* 19996 (holotype B).
Dendrobium kajewskii Ames in Journ. Arn. Arb. 13: 131 (1932). Type: Santa Cruz Islands, Vanikoro, *Kajewski* 638a (holotype AMES).

Stems clustered, 1–3 m tall. *Leaves* coriaceous, oblong, elliptic or ovate-elliptic, 4–12 cm long, 1.85–5.5 cm wide, unequally bilobed at obtuse or rounded apex, dark green; leaf bases sheathing. *Inflorescences* terminal, spreading or suberect, 30–66 cm long, rather densely 6–20 or more-flowered; peduncle 10–14 cm long. *Flowers* appearing contorted, yellow-green to yellow, more or less flushed with red-brown, lip yellow, veined with purple or brown; pedicel and ovary 2.5–4 cm long; sepals and petals have strongly undulate margins; dorsal sepal reflexed and sometimes spirally twisted, narrowly oblong-lanceolate, obtuse or subacute, 2–2.3 cm long; lateral sepals reflexed, obliquely oblong-lanceolate, acute, 2.3–2.4 cm long; petals linear-spathulate, acute or obtuse, 2.2–3 cm long, spirally twisted; lip trilobed, 2.1–2.35 cm long, 1.3–1.75 cm wide, lateral lobes narrowly elliptic, rounded or obtuse and spreading at apex, with verrucose venation and erose margins, mid-lobe recurved at apex, ovate, acute, with erose margins, callus of 3 keels, the middle one prominently raised at apex on base of mid-lobe; column 4–6 mm long. (See fig. 23, plate 8e).

DISTRIBUTION: Efate, Espiritu Santo and Pentecost. Also in New Guinea, Bougainville, the Solomon Islands and the Santa Cruz Islands.
HABITAT: Epiphytic on *Calophyllum inophyllum,* in secondary forest behind the beach, lowland forest and in old coconut plantations, sea level up to moderate altitudes.
COLLECTIONS: *Bregulla* 2 & 17 (K); *Cribb* 59 (K); *Cribb & Wheatley* 1 & 7 (K, PVNH); *Im Thurn* 344 (K); *MacKee* in RSNH 38290, 40231 & 41885 (P); *Vieillon* in RSNH 4508 (K, P); *Wheatley* 174 (K).

Dendrobium macranthum often grows with *D. conanthum* and may hybridise with it (see plate 8b).

19. D. gouldii *Reichb. f.* in Gard. Chron. 1867:901 (1867). Type: Solomon Islands, *J.G. Veitch* s.n. (holotype W!).
Dendrobium gouldii var. *acutum* Reichb. f., Xenia Orch. 2: 167, t.169 (1870). Type: as for *D. gouldii*.
Dendrobium undulatum var. *woodfordianum* Maid. in Proc. Linn. Soc. N.S.W. 24: 652 (1899). Type: Solomon Islands, *Woodford* s.n. (holotype NSW!).
Dendrobium woodfordianum (Maid.) Schltr. in Fedde, Rep. Sp. Nov. 1: 545 (1912).
Dendrobium imthurnii Rolfe in Kew Bull. 1912 & in Bot. Mag. 138, t. 8452. (1912). Type: Vanuatu, Efate, *Im Thurn* s.n. (holotype K!).

Stems clustered, 0.9–1.8 m tall, often slightly dilated in the middle. *Leaves* coriaceous, oblong-elliptic or elliptic, 12–17.5 cm long, 5–6.3 cm wide, obtuse or rounded at unequally bilobed apex; leaf bases sheathing. *Inflorescences* 30–70 cm long, 7–40 flowered. *Flowers* very variable in colour and size; sepals white, pale yellow or yellow with white, pale yellow, brown or violet petals; lip white to yellow, veined with purple or violet; pedicel and ovary 2.8–4.7 cm long; dorsal sepal recurved, linear-lanceolate, acute, 2–2.6 cm long; lateral sepals falcate, lanceolate, acute or acuminate, 2.2–2.6 cm long; mentum conical, 8–10 mm long, deflexed slightly at apex; petals linear-spathulate, obtuse, 2.7–4.0 cm long, one–three-times twisted; lip trilobed, 2.2–2.4 cm long, 1.3–1.5 cm wide, lateral lobes obliquely oblong, rounded in front, erose on margins, mid-lobe subspathulate, obtuse, apiculate, erose on margins, callus of 5 ridges, the central 3 prominent and raised at apex on centre of mid-lobe; column 6–8 mm long.

DISTRIBUTION: Efate, Erromango and Espiritu Santo. Also in New Ireland and the Solomon Islands.
HABITAT: Open park-like country, for example beside roads and in clearings, lowland up to moderate altitude (500 m).
COLLECTIONS: *Bregulla* 1 (K); *Im Thurn* s.n. (K); *Raynal* in RSNH 16235 (P).

This species is widely cultivated.

20. D. macranthum *A. Rich.* in Sert. Astrol.: 15, t.6. (1832). Type: Santa Cruz Islands, Vanikoro, *A. Richard* 39 (holotype P!).
Dendrobium arachnostachyum Reichb. f. in Gard. Chron. n. ser. 7: 334 (1877). Type: New Guinea, but more likely from a Pacific Island, *P. Veitch* s.n. (holotype W).
Dendrobium pseudotokai Kraenzl. in Bull. Soc. Bot. Fr. 76: 300 (1929). Type: Vanuatu, Efate, *Levat* s.n. (holotype P!).
Dendrobium tokai var. *crassinerve* Finet in Bull. Soc. Bot. Fr. 50: 381 (1903). Types: New Caledonia, *Thibault* s.n., *Germain* s.n. & *Balansa* 2387 (syntypes P!).

Stems clustered, up to 60 cm tall, slightly swollen in lower half and at base. *Leaves* distichous, coriaceous, oblong to elliptic, 7–12 cm long, 2.5–4.5 cm wide, unequally bilobed at apex; leaf bases sheathing. *Inflorescences* erect or suberect, 22–32 cm long, laxly 8–20-flowered. *Flowers* yellow, yellow-green or lime green; lip whitish, greenish or pale yellow in middle, with lilac calli and purple side veins; pedicel and ovary 2.5–5 cm long; dorsal sepal erect, linear-lanceolate, acute or acuminate, 2.4–3.8 cm long; lateral sepals recurved, falcate, obliquely linear-lanceolate, acuminate, 3–3.6 cm long; mentum narrowly conical, slightly

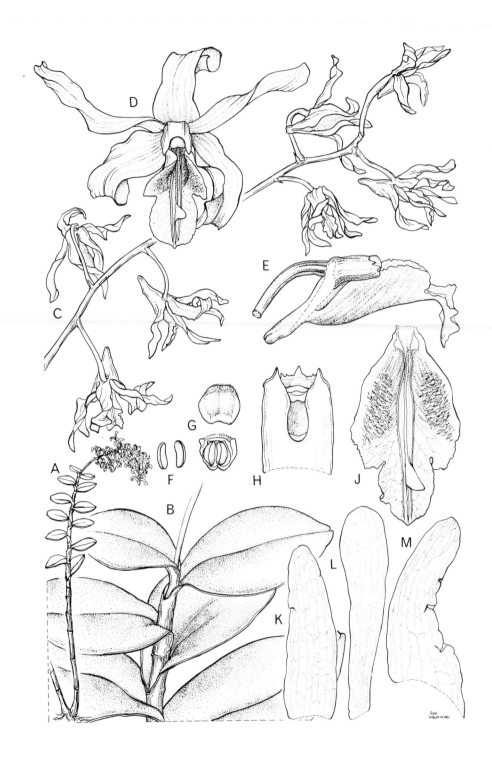

upcurved or decurved, 1–1.2 cm long; petals linear-subspathulate, acute, 3–4.5 cm long, not or once-twisted; lip trilobed in middle, porrect, 2.5–3.2 cm long, 1.2–1.3 cm wide, lateral lobes obliquely oblong, truncate in front, erose on margins, mid-lobe linear-lanceolate, acuminate, callus of 3 ridges, all slightly raised at apex on basal half of mid-lobe; column 6 mm long. (See plate 8a).

DISTRIBUTION: Anatom, Banks Islands (Vanua Lava), Efate, Epi, Erromango, Espiritu Santo and Pentecost. Also in the Santa Cruz Islands, New Caledonia, the Horn Islands and Samoa.
HABITAT: Strand and lowland forest, sea level to 500 m.
COLLECTIONS: *Baker* 111 (BM); *Bernardi* 13040 (G, P); *Boyd* s.n. (P); *Bregulla* 17 (K); *Cribb & A. Morrison* 1765 & 1774 (K); *Hallé* in RSNH 6296A (P); cult. *Lecoufle* (K); *McKee* 31247 & 31248 (P); *Raynal* in RSNH 16248 (P) & 16267 (K, P, PVNH); *de la Rüe* s.n. (P); *Slade* s.n. (K); *Wells* s.n. (K); *Wheatley* 355 (K, PVNH).

section RHOPALANTHE

21. D. goldfinchii *F. Muell.* in Wing's South Sci. Rec. 3: 4 (1883). Type: Solomon Islands, *Goldfinch* s.n. (holotype MEL).

Stems clustered, pendent or erect, strongly constricted below, swollen above, up to 50 cm high, may be zig-zag, with a reddish tinge; leaf bases sheathing. *Leaves* alternate, stiff, thick, bilaterally flattened, lanceolate-linear, 5–8 cm long, c. 0.5 cm wide, red to green. *Inflorescence* terminal; peduncle woody, up to 25 cm long. *Flowers* 1–4 from each node on peduncle, up to 20 in a raceme; pale yellow-green with a faint tinge of yellow; dorsal sepal lanceolate, 0.8 cm long; lateral sepals obliquely ovate, 1.0 cm long; mentum conical, upcurved, c. 1 cm long; petals linear, 0.8 cm long; lip obovate-cuneate, 1.1 cm long, 0.6 cm wide, with a central callus; column c. 3 mm long.

DISTRIBUTION: Banks Islands (Vanua Lava), Efate, Maewo and Pentecost. Also in Bougainville, the Solomon Islands, the Santa Cruz Islands and Samoa.
VERNACULAR NAME: Tuku.
HABITAT: Open habitats with plenty of light, lowland up to moderate altitudes (320 m).
COLLECTIONS: *Bregulla* 16 (K); *Cabalion* 1028 (PVNH) & 1236 (P); *Morrison* in RBG Kew 110 (K); *Wheatley* 171 & 319 (K, PVNH).

section GRASTIDIUM

22. D. biflorum (*G. Forst.*) *Sw.* in Nov. Act. Soc. Sc. Upsal. 6: 84 (1799).
Epidendrum biflorum G. Forst., Fl. Ins. Austr. Prodr.: 60 (1786). Type: Tahiti, *G. Forster* s.n. (holotype BM!, isotype K!).
Dendrobium vanikorense Ames in Journ. Arn. Arb. 13: 134 (1932). Type: Santa Cruz Islands, Vanikoro, *Kajewski* 663 (holotype AMES).

Fig. 23. *Dendrobium conanthum.* **A**, habit × $\frac{1}{10}$; **B**, leaves × $\frac{2}{3}$; **C**, inflorescence × $\frac{2}{3}$; **D**, flower ×$\frac{2}{3}$; **E**, column and lip×2; **F**, pollinia×4; **G**, anther cap×4; **H**, column×4; **J**, lip×2; **K**, dorsal sepal×2; **L**, petal×2; **M**, lateral sepal×2. All drawn from *Wickison* 69 (Kew spirit no. 50907) by Sue Wickison.

Stems clustered, pendulous, up to 1 m long, 1.5–2 mm wide; internodes 1–2.5 cm long. *Leaves* linear-lanceolate, tapering gradually to an acute apex, 8–11 cm long, 3 mm wide; sheaths persistent. *Flowers* lateral, in pairs, white, creamy or yellow, may be tinged with purple; pedicels c. 13 mm long; sepals and petals lanceolate, tapering to a long filiform apex, 1.2–2.8 cm long; mentum conical, obtuse; lip trilobed, 0.9–1.2 cm long, 0.4–0.5 cm wide, lateral lobes erect, crescent-shaped, mid-lobe narrowly triangular, papillose, elongated and filiform at apex, disc with a longitudinal callus; column c. 3 mm long, foot c. 4.5 mm long.

DISTRIBUTION: Ambae, Anatom, Efate, Erromango and Pentecost. Also in Bougainville, the Solomon Islands, the Santa Cruz Islands, Fiji, Samoa and Tahiti.
HABITAT: *Agathis* and other forest in plenty of light, 100–500 m.
COLLECTIONS: *Bourdy* 220 (P, PVNH); *Bregulla* 24 (K); *Cheesman* s.n. & A73 (BM); *Green* in RSNH 1310 (K, P, PVNH); *Morat* 6402 (P); *Raynal* in RSNH 16234 (K, P); *de la Rüe* s.n. (P); *Schmid* 3789 (P); *Chew Wee-Lek* in RSNH 142 (K, P); *Wheatley* 88, 293 & 293 (K, PVNH).

Guillaumin identified the collections of *Cheesman* from Vanuatu as *Dendrobium acuminatissimum* Lindley, however we consider that these specimens are *D. biflorum*.

D. biflorum is similar to *D. meliodorum* Schltr. from New Guinea, but *D. meliodorum* has slightly longer, narrower leaves.

23. D. greenianum *Cribb & B. Lewis* in Orchid Rev. in press (1989). Type: Vanuatu, Pentecost, *Wheatley* 198A (holotype K!).

Stem pendent, leafy up to c. 60 cm long, 2–2.5 mm wide, slightly bilaterally compressed, covered by sheathing leaf bases; internodes 1.2–2.2 cm long. *Leaves* distichous, twisted at the base to lie in one plane, coriaceous, lanceolate, minutely unequally bilobed at the subacute apex, 5–6.8 cm long, 0.7–1.4 cm wide; leaf bases 1–2 cm long. *Inflorescence* emerging through the leaf bases opposite the leaves, 2-flowered at each node, sessile; bracts ovate, acute, 2 mm long. *Flowers* white or creamy white; pedicel 2–3 mm long; dorsal sepal oblong-subspathulate, obtuse, 7–7.5 mm long, 3 mm wide; lateral sepals obliquely oblong, obtuse, 7.5–8 mm long, 4 mm wide, dilated at base to form a saccate mentum c. 3 mm long; petals spathulate, obtuse, 6 mm long, 2 mm wide; lip obscurely trilobed in the middle, elliptic, obtuse, 7 mm long, 3.5 mm wide, with transverse rugulose ridges on the surface and a longitudinal low raised central callus almost to the apex, lateral lobes very narrow, acute, midlobe ovate; column 2 mm long. (See fig. 24).

DISTRIBUTION: Pentecost.
HABITAT: Epiphyic on *Eugenia* in light shade, 310 m.
COLLECTION: *Wheatley* 198A (K).

24. D. kietaense *Schltr.* in Fedde, Rep. Sp. Nov., Beih. 1: 614 (1912). Type: Bougainville, *Rechinger* 4824 (holotype B).

Fig. 24. *Dendrobium greenianum*. **A**, habit × ⅔; **B**, lip × 6; **C**, flower × 4; **D**, dorsal sepal ×6; **E**, petal × 6; **F**, lateral sepal × 6; **G**, anther cap × 10; **H**, pollinia × 10; **J**, column × 10. Drawn from *Wheatley* 198A by Sue Wickison.

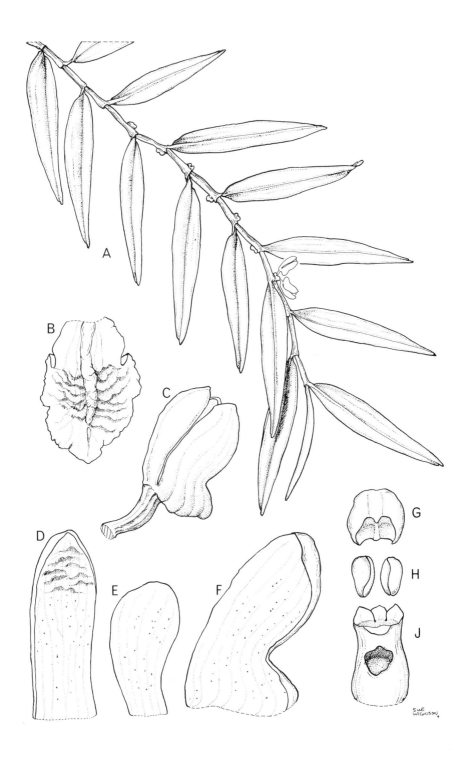

SUE
WICKISON

119

Stems clustered, pendulous, up to 1.5 m long, c. 5 mm wide. *Leaves* lanceolate, 9–17 cm long, 0.8–1.4 cm wide, unequally bilobed at apex, light green; sheaths persistent. *Flowers* lateral, in pairs, fleshy, not opening widely, greenish-yellow to white; pedicels 7–10 mm long; sepals and petals 1.5 cm long; dorsal sepal oblong; lateral sepals falcate-oblong; mentum obtuse; petals elliptic; lip trilobed, pandurate, keeled, c. 1 cm long, 4 mm wide, lateral lobes crescent-shaped; column 3.5 mm long, with a 2 mm long foot.

DISTRIBUTION: Ambae and Banks Islands (Vanua Lava). Also in Bougainville and the Solomon Islands.
HABITAT: Mixed evergreen rain forest, up to 500 m.
COLLECTIONS: *Cabalion* 455 (P); *Veillon* 5543 (P); *Wheatley* 85 (K, PVNH).

25. D. involutum *Lindley* in Journ. Linn. Soc. 3: 15 (1859). Type: Society Islands, *Matthews* s.n. (holotype K!).
Dendrobium cheesmanae Guillaumin in Bull. Soc. Bot. Fr. 103: 280 (1956); **synon. nov.** Type: Vanuatu, Anatom, *L.E. Cheesman* A22 (holotype BM!).

Stems clustered, pendent, up to 80 cm long; mature stems clothed with fibres from old leaf sheaths; internodes 1–1.5 cm wide. *Leaves* lanceolate, 4.5–7 cm long, 0.7–2.0 cm wide, unequally bilobed at apex, thick; leaf bases sheathing. *Flowers* lateral in pairs, creamy yellow-orange; pedicels c. 1.0 cm long; sepals and petals lanceolate, c. 1.2 cm long; mentum obtuse, upcurved, c. 0.5 cm long; lip reflexed, especially at apex, trilobed, c. 1 cm long, 0.7 cm wide, creamy-yellow, with midrib of orange-yellow and transversely radiating purplish-red lines, lateral lobes erect, triangular, mid-lobe triangular, rugulose, with erose margins, callus central extending to just beyond isthmus of lateral lobes; column 3 mm long, with a 3 mm long foot.

DISTRIBUTION: Ambae, Anatom, Banks Islands (Vanua Lava), Efate, Erromango, Espiritu Santo and Pentecost. Also in the Santa Cruz Islands and the Society Islands.
HABITAT: Rain forest, epiphyte on *Calophyllum*, sea level to 500 m.
COLLECTIONS: *Bourdy* 155 (P) & 175 (P, PVNH); *Cabalion* 2717 (P); *Cribb & Wheatley* 55 & 126 (K, PVNH); *Green* in RSNH 1295 & 1357 (K); *Wheatley* 78, 189 & 325 (K, PVNH).

In the collections which we have studied the shape of the leaves is variable and futher study may show that several distinct species are included in this concept.

26. D. sladei *J.J. Wood & Cribb* in Orchid Rev. 90(1059): 14 (1982). Type: Vanuatu, Efate, *Im Thurn* 330 (holotype K!).

Stems clustered, pendent, terete, up to 2 m long, covered by sheathing leaf bases. *Leaves* distichous, lanceolate, 5–11 cm long, 1.5–2.3 cm wide, with a minutely unequally bilobed apex. *Inflorescences* lateral; peduncle c. 6 mm long. *Flowers* in pairs, cream-straw yellow, scented; pedicels 1.2 cm long; sepals lanceolate, acute, cucullate; lateral sepals falcate, 2.5–3 cm long, fused at the base to form a mentum; mentum incurved, conical, 0.5–0.6 cm long; petals ligulate-lanceolate, 2.5 cm long; lip trilobed, strongly recurved particularly towards the apex, 1.1 cm long, 0.7 cm wide, lateral lobes erect, narrowly triangular, acute, minutely erose, purple-red, mid-lobe triangular-ovate, acuminate, erose-undulate at margin, disc almost entirely rugose-tuberculate, glabrous at apex of lobes, callus a low, median ridge extending from the base and merging just beyond the isthmus of the lateral lobes into the papillose median nerve; column 4 mm long, cream, with an incurved, 5–6 mm long foot, with a large orange blotch. (See plate 5d).

DISTRIBUTION: Banks Islands (Vanua Lava), Efate, Erromango, Espirito Santo and Pentecost. Also in Fiji.

HABITAT: Coastal forest, sea level to 400 m.

COLLECTIONS: *Bourdy* 155 (K, NOU); *Bregulla* 22 (K); *Cabalion* 1369 (PVNH); *Cribb & A. Morrison* 1794 (K); *Cribb & Wheatley* 17 (K, PVNH); *Green* in RSNH 1048, 1295 & 1357 (K); *Im Thurn* 329 & 330 (K); *Veillon* 4404 (P); *Webster & Bennett* s.n. (P); *Wheatley* 226 & 417 (K, PVNH).

section DICHOPUS

27. D. insigne (*Blume*) *Reichb. f. ex Miq.*, Fl. Ind. Bat. 3: 640 (1859).
Dichopus insignis Blume in Mus. Bot. Lugd. Bat. 2: 176 (1856). Type: New Guinea, *Zippelius* (holotype L).
Dendrobium gazellae Kraenzl. in Engler, Bot. Jahrb. 7: 436 (1886). Type: New Guinea, *Dr. Naumann* s.n. (holotype B).
Dendrobium lyperanthiflorum Kraenzl. in Oest. Bot. Zeitschr. 44: 334 (1894). Type: New Britain, *W. Micholitz* s.n. (holotype W).
Dendrobium obcuneatum Bail. in Queensl. Agric. Journ. 17: 231 (1906). Type: New Guinea, *Rev. Copeland King* s.n. (holotype BRI, isotype K!).
Dendrobium pentactis Kraenzl. in Engl., Pflanzenr. Orch. Mon.- Dendr.: 200 (1910). Type: New Guinea, *Beccari* s.n. (holotype FI).

Stems clustered, up to 80 cm long, not swollen at the base. *Leaves* distichous, broadly ovate, thick, 6–9 cm long, 1.5–4.0 cm wide, becoming smaller towards the apex of the stem (1.5–2.0 cm long, 0.7–1.0 cm wide), unequally bilobed at apex; leaf bases sheathing, their distal ends may appear dark in dried specimens giving the stem a ringed appearance. *Flowers* lateral, single or in pairs, yellow with a reddish tinge to dark red; pedicels 0.5–1.5 cm long; dorsal sepal linear, 2.5 cm long; lateral sepals linear-falcate, 2.5 cm long; mentum obtuse; petals linear-falcate, 2.2 cm long; lip trilobed c. 1 cm long, 0.4 cm wide with a central callus, lateral lobes fimbriate, mid-lobe triangular; column c. 0.7 cm long, with a movable appendage on the lower margin of the stigma.

DISTRIBUTION: Efate. Also in New Guinea, New Britain, New Ireland, the Solomon Islands and Saibai Island (Australia).

HABITAT: No information from Vanuatu. In New Guinea this species frequently occurs near the coast, especially where there is exposure to the sun.

COLLECTION: No specimens seen, (Guillaumin, 1927 & 1948).

Guillaumin cites two specimens, *Im Thurn* 329 & 330, from Efate as being *Dendrobium insigne*, however we have not seen these collections and the description is from New Guinea specimens.

We have followed Schlechter in listing the synonyms shown above but M.A. Clements (pers. comm.) considers that some of the names shown as synonyms refer to good species.

section MONANTHOS

28. D. bilobum *Lindley* in Hook., Lond. Journ. Bot. 2: 236 (1843). Type: New Guinea, *Hinds* s.n. (holotype K!).
Cadetia biloba (Lindley) Blume in Mus. Bot. Lugd. Bat. 1: 30 (1849).
Callista biloba (Lindley) Kuntze, Rev. Gen. Pl. 2: 654 (1891).

Stems clustered, erect or pendent, 20–40 cm high; internodes 1.3–2 cm long. *Leaves* alternate, lanceolate-oblong, 4–7 cm long, 4–5 mm wide, bilobed at apex, bright green; leaf sheaths persistent. *Flowers* lateral, solitary, non resupinate; cream, pale yellow or yellowish green; pedicel 2–3 mm long; dorsal sepal ovate, 4–5 mm long; lateral sepals obliquely triangular, 6.5 mm long; mentum obtuse, slightly recurved; petals linear-lanceolate, 3.5 mm long; lip trilobed, fleshy, 5 mm long, 3 mm wide, maroon or maroon and pale yellow, lateral lobes erect, semicircular, mid-lobe ovate, apex fleshy, verrucose; column c. 2 mm long, cream; foot c. 9 mm long. *Fruit* 10 mm long, 5.5 mm wide.

DISTRIBUTION: Espiritu Santo. Also in New Guinea, the Solomon Islands, New Caledonia and Fiji.
VERNACULAR NAME: Tuku.
HABITAT: Montane forest, 900 m.
COLLECTION: *Cribb & Wheatley* 40 (sterile) (K, PVNH).

S. Wickison (pers. comm.) notes that in the Solomon Islands the leaves of this species are smaller and more numerous at higher altitudes (above 320 m).

50. **DIPLOCAULOBIUM** (Reichenbach f.) Kraenzlin

Pseudobulbs clustered, slender, short or long, but never very long, somewhat swollen in the lower portion only. *Leaves* solitary, terminal. *Flowers* last for less than a day, borne on pedicels from within a large bract at the apex of the pseudobulb, solitary or occasionally 2; sepals and petals very slender, often filiform; bases of lateral sepals rather broad and forming with the column-foot a distinct mentum; lip trilobed, lateral lobes not very large, disc with keels; column with a distinct, but not very long foot; pollinia 4.
A genus of about 100 species centred on New Guinea and extending to Malaysia in the west and Australia and the Pacific Islands in the east. A new genus record for Vanuatu, a single species being recorded.

D. ouhinnae (*Schltr.*) *Kraenzl.* in Engl., Planzenr. Orch. Mon.-Dendr.: 337 (1910). *Dendrobium ouhinnae* Schltr. in Engler, Bot. Jahrb. 39: 68 (1906). Type: New Caledonia, *Schlechter* 15626 (isotype P!).

Epiphytic. Pseudobulbs clustered, up to 30 cm long, yellow, bases slightly swollen, covered by bracts. *Leaf* solitary, apical, oblong, c. 7.5 cm long, 0.8 cm wide, bilobed at apex. *Bract* 2–2.5 cm long. *Flower* single, terminal, white with yellow tips maturing purple; pedicel c. 5 cm long; sepals and petals filiform, 3.2–3.6 cm long; lateral sepals form a mentum, c. 4 mm long; lip trilobed, with a median callus, 9 mm long, 4 mm wide, lateral lobes semiovate, midlobe clawed, expanding into an ovate lamina. *Fruit* 2.5 cm long, 0.7 cm wide.

DISTRIBUTION: Ambae, Efate and Espiritu Santo. Also in New Caledonia.

HABITAT: Ridge-top forest, 600–800 m.
COLLECTIONS: *Cribb & Wheatley* 62 (K, PVNH); *MacKee* 32089 & 37353 (P); *Suprin* 276 (P); *Wheatley* 98 (K, PVNH).

51. FLICKINGERIA Hawkes

Epiphytic. Stems branching, each branch of stem terminates in a pseudobulb, which bears a solitary leaf at its apex, new branches commence from the bases of the pseudobulbs. *Flowers* ephemeral, from apex of pseudobulb, borne on individual pedicels from a group of bracts; lip with or without an apical fringe; pollinia 4.

A genus of 50 to 65 species from tropical Asia to Australia. A single species in Vanuatu.

F. comata *(Blume) Hawkes* in Orquidea 27: 301 (1965).
Desmotrichum comatum Blume, Bijdr.: 230 (1825). Type: Java, *Blume* s.n. (holotype L, isotype P!).
Dendrobium comatum (Blume) Lindley, Gen. Sp. Orch. Pl.: 76 (1830).
Dendrobium criniferum Lindley in Bot. Reg. 30: misc. 41 (1844). Type: Sri Lanka, *Power* s.n. (holotype K!).
Ephemerantha comata (Blume) P.F. Hunt & Summerhayes in Taxon 10: 102 (1961).

Plant erect or pendent, up to 1 m long. *Pseudobulbs* becoming progressively smaller the farther they are from the base of the branching stem. *Stems* yellowish. *Leaves* also becoming progressively smaller the farther they are from the base of the stem, elliptic, 6–22 cm long, 3–10 cm wide. *Flowers* 1–8, opening widely, cream to pale yellow with or without pale purple spots, with sidelobes of lip purple and apex of midlobe densely fringed with hairs which vary from cream to yellow to pale purple; sepals and petals 1–1.5 cm long; lip trilobed, 1.3–1.9 cm long, 6–8 mm wide when flattened, lateral lobes transversely oblong, midlobe tapered from the base to a blunt apex, with crenate margins, distal third fringed with thick hairs, c. 5 mm long, disc with two parallel, crenate keels which extend to the apex of the midlobe; column c. 4.5 mm long, foot c. 6 mm long. (See plate 7f).

DISTRIBUTION: Banks Islands (Vanua Lava), Efate, Espiritu Santo and Pentecost. Widely distributed from Sri Lanka, Indonesia and the Malay peninsula to Sumatra, New Guinea, the Solomon Islands, New Caledonia, Samoa and Australia.
HABITAT: Mangroves and rain forest, sea level to 480 m.
COLLECTIONS: *Cribb & Wheatley* 121 (K, PVNH); *Green* in RSNH 13463 (K); *McKee* 42523 (P); *Wheatley* 300 & 353 (K, PVNH).

McKee 42523 (P) has much narrower leaves (6.5–8 cm long, 1.3–1.7 cm wide) than is typical for this species.

52. BULBOPHYLLUM Thouars

Epiphytic. Pseudobulbs 1-leaved, clustered or spaced on creeping rhizome. *Flowers* solitary or many, small to medium-sized; sepals and petals free; lateral sepals united at base to column-foot to form a short mentum; lip entire, fleshy, usually hinged to column-foot; column short usually with a short column-foot; pollinia 2 or 4.

A large cosmopolitan genus of about 1000 species. *Bulbophyllum* is the second largest genus in Vanuatu, after *Dendrobium*, with 17 species, 8 being newly recorded there, namely, *B. atrorubens, B. membranaceum, B. microrhombos, B. minutipetalum, B. polypodioides, B. stenophyllum, B. streptosepalum* and *B. sp. nov.* The species have been arranged in 14 sections following Schlechter (1912).

1. Flowers solitary, or borne one at a time 2
   Flowers two or more 13
2. Rhizome dichotomising at nodes which bear pseudobulbs
   ... ... **9. B. sp. nov.**
   Rhizomes not as above   3
3. Lateral sepals c. 6 cm long, with spirally entwined apices, yellow, striped
   dark maroon   ... ... **7. B. streptosepalum**
   Lateral sepals less than 3 cm long, not entwined 4
4. Peduncle 28–60 cm long; lip fleshy, more than 2 cm long
   ... ... **3. B. longiscapum**
   Peduncle less than 15 cm long; lip less than 1.5 cm long 5
5. Leaves petiolate   6
   Leaves without a petiole 8
6. Sepals white-creamy yellow; lip rhomboid, white-yellow, with apex of lateral
   lobes brownish   ... ... **5. B. rhomboglossum**
   Sepals purple or pale straw yellow with purple lines basally;   lip elliptical,
   yellow with a purple apex or yellow covered in purple-brown papillae
   7
7. Leaves 9–16 cm long, 1.5–2.0 cm wide; sepals 1.6–2.0 cm long
   ... ... **1. B. stenophyllum**
   Leaves 3.5–6 cm long, 0.5–0.6 cm wide; sepals 1.0–1.3 cm long
   ... ... **2. B. microrhombos**
8. Pseudobulbs at right angles to the rhizome   9
   Pseudobulbs appressed to the rhizome 11
9. Leaves ovate, 2–4 cm long, (0.7)1–1.5 cm wide; pseudobulbs 2.5–6 cm apart
   ... ... **8. B. membranaceum**
   Leaves oblong, 6–15 cm long, 1–3 cm wide; pseudobulbs c. 0.5 cm apart or
   less 10
10. Peduncle c. 3 cm long; petals c. 0.7 mm long, obliquely mucronate; lip square
    with rounded edges ... ... **4. B. minutipetalum**
    Peduncle 7–10 cm long; petals c. 3 mm long, linear; lip tongue-shaped
    ... ... **6. B. samoanum**
11. Peduncle 3–7 cm long, extending beyond leaves   ... ...**12. B. saviense**
    Peduncle less than 1.5 cm long, not extending beyond leaves 12
12. Leaves 3.5–5 cm long; sepals purple with green apices; lip tongue-shaped,
    densely covered in papillae   ... ... **10. B. betchei**
    Leaves 1.4–2.0 cm long; sepals white to white-green; lip oblong with a bifid
    basal callus ... ... **11. B. neocaledonicum**
13. Inflorescence subumbellate   ... ... **17. B. longiflorum**
    Inflorescence not subumbellate 14
14. Leaves obovate; inflorescence with flowers in a dense globose head
    ... ... **16. B. atrorubens**
    Leaves oblong to lanceolate; inflorescence racemose 15

15. Flowers 10–20; lip bright yellow ... ...   **15. B. polypodioides**
    Flowers 3–6; lip greenish or white 16
16. Sepals lanceolate-acuminate, 1.2 cm long, white ... ...   **13. B. pensile**
    Sepals ovate, 4 mm long, yellow with maroon stripes   ... ...   **14. B. setipes**

section COELOCHILUS

1. B. stenophyllum *Schltr.* in Fedde, Rep. Sp. Nov., Beih. 1: 719 (1912). Type: New Guinea, *Schlechter* 18521 (holotype B).

Rhizome creeping. *Pseudobulbs* 1–2.5 cm tall, c. 0.5 cm wide, 0.5–4.0 cm apart. *Leaf* lanceolate-ovate, 9–16 cm long, 1.5–2.0 cm wide; petiole c. 1.5 cm long. *Inflorescence* arising from rhizome, 3–10 cm long. *Flower* apical, solitary; sepals lanceolate, 1.6–2.0 cm long, pale straw-yellow with a few purple lines at the base; petals lanceolate, falcate, c. 3 mm long, pale straw-yellow with a central purple line and purple blotches at the base; lip ovate-elliptic, convex, c. 1.3 cm long, yellow, densely covered in purple-brown papillae except at the apex; column footless.

DISTRIBUTION: Ambae, Efate and Pentecost. Also in New Guinea, Bougainville, the Solomon Islands and the Horn Islands.
HABITAT: Montane forest on volcanic soils, 480–980 m.
COLLECTIONS: *Green* 1060 (K); *Wheatley* 69 & 109 (K).

Bulbophyllum trachyglossum Schltr. and *B. chrysoglossum* Schltr. from New Guinea are similar, but *B. stenophyllym* is distinguished from them by its lanceolate, falcate petals.

2. B. microrhombos *Schltr.* in Fedde, Rep. Sp. Nov., Beih. 1: 719 (1912). Type: New Guinea, *Schlechter* 19757 (holotype B).

Rhizome creeping. *Pseudobulbs* c. 1 cm high, 0.5–0.6 cm wide, thick, 0.5–1.0 cm apart. *Leaf* lanceolate-ovate, 3.5–6 cm long. *Inflorescence* arising from the rhizome, 4–6 cm long. *Flower* apical, solitary; sepals lanceolate, 1–1.3 cm long, purple with 3 darker veins and slightly paler margins; petals reduced, obliquely rhombic, c. 1.25 cm long; lip oblong-elliptic, c. 9 mm long, convex, papillose, yellow-purple at base, dark purple at tip, yellow on underside; column 2.5 mm long, yellow-green.

DISTRIBUTION: Ambae. Also in the Solomon Islands and New Guinea.
HABITAT: Mixed evergreen rain forest, 500–1100 m.
COLLECTIONS: *Wheatley* 36 & 95 (K, PVNH).

section DIALEIPANTHE

3. B. longiscapum *Rolfe* in Kew Bull. 1896: 45 (1896). Type: Fiji, *Yeoward* s.n. (holotype K!).
B. praealtum Kraenzl. in Notizbl. Bot. Gart. Berlin 5: 109 (1909). Type: Samoa, *Vaupel* 322 (holotype B).

Rhizome thick and woody. *Pseudobulbs* c. 2 cm high, 1–1.5 cm wide, 4–5 cm apart. *Leaf* oblong-lanceolate, 11–21 cm long, c. 2.8 cm wide, mid-green with a sheen; petiole 2–4 cm long. *Inflorescence* arising from the base of the pseudobulb;

peduncle 28–60 cm long. *Flowers* apical, opening one at a time, with an unpleasant scent; sepals lanceolate, c. 3.5 cm long, lemon-green with purple lines at the base; petals minute, mucronate, c. 4 mm long, yellow-green to purple; lip very fleshy, tongue-shaped, 2.3–3.0 cm long, maroon, with a yellow apex; column white; anther-cap yellow. (See fig. 25, plate 7a).

DISTRIBUTION: Efate, Espiritu Santo, Pentecost and Tanna. Also in Solomon Islands, the Mariana Islands, the Horn Islands, Fiji and Samoa.
HABITAT: Lowland and mountain forest, 310–600 m.
COLLECTIONS: *Bregulla* 10 (K); *Cribb & A. Morrison* 1816 (K); *Cribb & Wheatley* 71 & 99 (K, PVNH); *McKee* 32756, 34543 & 36380 (P); *Morat* 5918 (P); *Wheatley* 197 (K, PVNH).

section PELTOPUS

4. B. minutipetalum *Schltr.* in Fedde, Rep. Sp. Nov., Beih. 1: 761 (1912). Type: New Guinea, *Schlechter* 18535 (holotype B).

Rhizome creeping, bearing many adventitious roots. *Pseudobulbs* 0.7–1.5 cm tall, c. 0.5 cm wide, c. 0.5 cm apart. *Leaf* oblong, 6–8 cm long, 1.0–1.4 cm wide. *Inflorescence* arising from the base of the pseudobulb; peduncle c. 3 cm long. *Flower* apical, solitary, yellow; sepals lanceolate, elongate-acuminate, c. 1 cm long, with minutely ciliate margins; petals minute, obliquely mucronate, c. 0.7 mm long; lip fleshy, square with rounded edges, c. 4 mm long, arched in profile, slightly papillose, dark yellow; column c. 2 mm long. (See fig. 26).

DISTRIBUTION: Espiritu Santo. Also in New Guinea.
HABITAT: Ridge-top forest, 1550 m.
COLLECTION: *Cribb & Wheatley* 87 (K, PVNH).

section BRACHYPUS

5. B. rhomboglossum *Schltr.* in Fedde, Rep. Sp. Nov., Beih. 1: 767 (1913). Type: New Guinea, *Schlechter* 20364 (holotype B).

Rhizome covered in adventitious roots. *Pseudobulbs* conical, 1.5–2 cm high, c. 7 mm wide, clustered. *Leaf* ovate-lanceolate, 6–11 cm long, 1.5–3 cm wide; petiole 1–1.5 cm long. *Inflorescence* from base of pseudobulb; peduncle c. 2 cm long. *Flower* apical, solitary, white to creamy-yellow; sepals ovate, c. 8 mm long; petals linear, c. 5 mm long; lip rhombic, 8–9 mm long, with lateral lobes pronounced and reflexed, brownish at apex; column c. 6 mm long, pale yellow; anther brownish.

DISTRIBUTION: Epi and Espiritu Santo. Also in New Guinea.
HABITAT: Cloud forest, 900–1000 m.
COLLECTIONS: *Cribb & Wheatley* 46 (K, PVNH); *Maunder* s.n. (K).

Fig. 25. *Bulbophyllum longiscapum.* **A,** habit × ⅔; **B,** dorsal sepal × 2; **C,** petal × 4; **D,** lip and column × 3; **E,** anther cap × 6; **F,** flower × ⅔; **G,** pollinia × 6; **H,** lateral sepal × 2. **A** drawn from *Wickison* 51; **B–H** from *Bregulla* 10 (Kew spirit no. 39067) by Sue Wickison.

section PAPULIPETALUM

6. B. samoanum *Schltr.* in Fedde, Rep. Sp. Nov. 9: 107 (1911). Type: Samoa, *Vaupel* 546 (holotype B).
B. christophersenii L.O. Williams in Bot. Mus. Leafl. Harv. Univ. 7: 143 (1939). Type: Samoa, *E. Hume* 2297 (holotype AMES, isotypes K!, P!).

Rhizome creeping. *Pseudobulbs* c. 7 mm high, 4 mm wide, clustered. *Leaf* oblong, 7–15 cm long, 1.5–3 cm long. *Inflorescence* arising from the base of the pseudobulb; peduncle 7–10 cm long. *Flower* apical, solitary, yellow-green tinged with purple or red, with a white lip; sepals lanceolate c. 1 cm long; petals linear, minute, c. 3 mm long, with ciliate margins; lip tongue-shaped, 3–4 mm long; column c. 3 mm long, white. (See fig. 27).

DISTRIBUTION: Anatom, Efate and Espiritu Santo. Also in Fiji and Samoa.
HABITAT: *Metrosideros* forest and montane forest, 420–1200 m.
COLLECTIONS: *Bregulla* 3 (K); *Cabalion* 2760 (P); *Raynal* in RSNH 15999 & 16151 (P); *Wheatley* 10 (K, PVNH).

section EPHIPPIUM

7. B. streptosepalum *Schltr.* in Fedde, Rep. Sp. Nov., Beih. 1: 779 (1913). Type: New Guinea, *Schlechter* 19326 (holotype B).

Rhizome creeping, ascending. *Pseudobulbs* 0.7–1.0 cm high, c. 0.5 cm wide, 1–3 cm apart. *Leaf* oblong-elliptic, obtuse, 2–3 cm long, 1.2–1.6 cm wide. *Inflorescence* arising from the rhizome; peduncle 12–17 cm long. *Flower* apical, solitary; sepals yellow striped dark maroon; dorsal sepal ovate, concave c. 1 cm long; lateral sepals lanceolate, c. 6 cm long, twisted; petals small, trifid at apex, c. 3.5 mm long; lip tongue-shaped, c. 2 mm long, fleshy, with hairy margins; column c. 2 mm long. (See fig. 26, plate 7e).

DISTRIBUTION: Espiritu Santo. Also in New Guinea.
HABITAT: Epiphytic on *Casuarina* trunks in riverine forest, c. 350 m.
COLLECTION: *Cribb & Wheatley* 120 (K, PVNH).

section SCYPHOSEPALUM

8. B. membranaceum *Teijsm. & Binnend.* in Nat. Tijdschr. Ned. Ind. 3: 397 (1855). Type: Java, *Teijsmann & Binnendijk* s.n. (holotype BO).

Pseudobulbs ovoid, 6–7 mm tall, 5 mm wide, 2.5–6 cm apart. *Leaves* ovate, 2–4 cm tall, (0.7)1–1.5 cm wide. *Inflorescence* arising from the base of the pseudobulb or from the rhizome, 2 mm long. *Flower* apical, solitary, not opening widely, cream to white; sepals ovate, 4 mm long, lateral sepals broader than dorsal sepal; petals

Fig. 26. *Bulbophyllum streptosepalum*. **A**, habit × ⅔; **B**, flower × 3; **C**, dorsal sepal × 6; **D**, lip × 14; **E**, petal × 14; **F**, side view lip × 14; **G**, column × 10. *B. minutipetalum*; **H**, habit × ⅔; **J**, flower × 3; **K**, lip × 10; **L**, column × 14; **M**, column and petals × 10; **N**, dorsal sepal × 4; **O**, lateral sepal × 4; **P**, lip × 10. A–G drawn from *Cribb & Wheatley* 120 (Kew spirit no. 53173); **H–P** from *Cribb & Wheatley* 87 (Kew spirit no. 53176). All drawn by Sue Wickison.

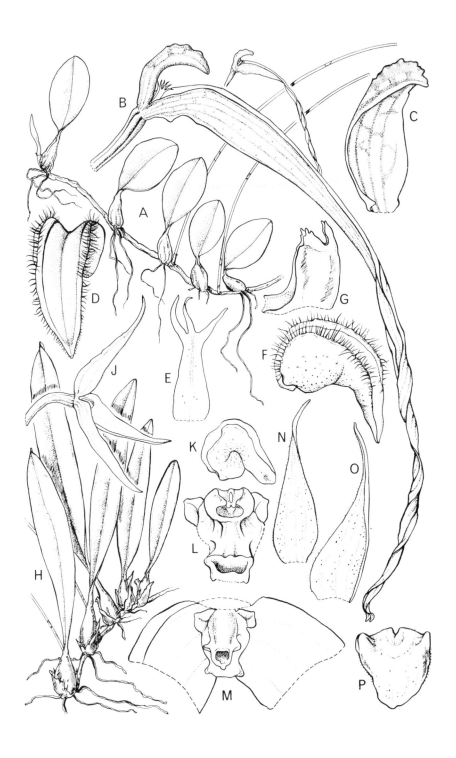

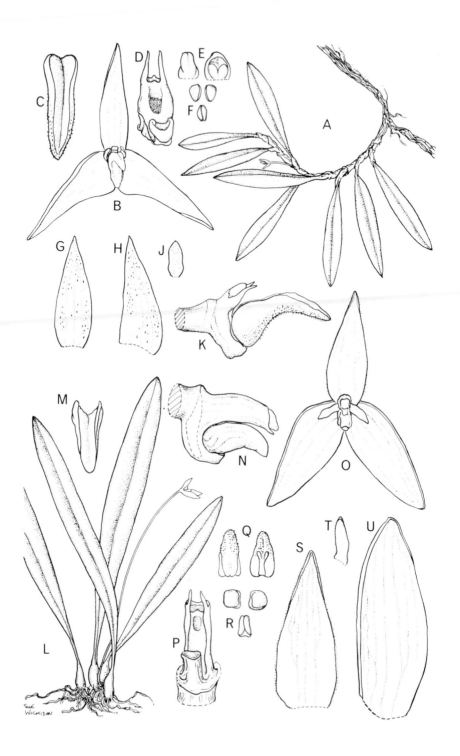

small, mucronate, 1.5 mm long; lip tongue-shaped, papillose, 1.5 mm long, reflexed at apex; column 1.5 mm long.

DISTRIBUTION: Erromango and Tanna. Widely distributed from the Malay peninsula to Samoa.

HABITAT: Rain forest, 500–900 m.

COLLECTIONS: *Bernardi* 13346 (P); *Raynal* in RSNH 16000 (P); *Schmid* s.n. (P).

In the Malay peninsula this species has purple flowers and a yellow lip.

section LEPTOPUS

9. B. sp. nov.

Rhizome creeping, pendent, dichotomising at nodes which bear pseudobulbs. *Pseudobulbs* conical, 1.5–2 cm tall, c. 2 mm wide, pale yellow; 4–7 cm apart. *Leaf* lanceolate, acute, 5–6.5 cm long, 7–10 mm wide, bright mid-green, shiny above, paler below. *Inflorescence* 1-flowered; peduncles short. *Flowers* not seen.

DISTRIBUTION: Ambae, Anatom and Espiritu Santo.

HABITAT: Submontane forest, 150–700 m.

COLLECTIONS: *Bourdy* 410 (K, P); *Cribb & Wheatley* 36 (K, PVNH); *Schmid* 3982 (P); *Wheatley* 75 (K, PVNH).

All the collections which we have seen are sterile.

section FRUCTICOLA

10. B. betchei *F. Muell.* in Wing's South Sci. Rec. 1: 173. (1881) Type: Samoa, *Betche* 261 (holotype MEL).

B. finetianum Schltr. in Engler, Bot. Jahrb. 39: 83 (1906). Type: New Caledonia, *Schlechter* 15416 (holotype B, isotypes K!, P!, Z).

B. atroviolaceum Fleischm. & Rech. in Denkschr. Akad. Wien, Math.-Nat. 85: 261 (1910). Type: Samoa, *Rechinger* 1824 (holotype W).

B. ponapense Schltr. in Engler, Bot. Jahrb. 56: 484 (1921). Type: Caroline Islands, *Ledermann* 13447 (holotype B).

Rhizome creeping, bearing many adventitious roots and cataphylls. *Pseudobulbs* cylindrical, appressed to rhizome, c. 5 mm long, 1–2 mm wide, 3–15 mm apart. *Leaf* oblong-lanceolate, 3.5–5 cm long, 0.5–1.0 cm wide, yellow-green. *Inflorescence* arising from the rhizome, thin, 0.5–1.0 cm long. *Flower* apical, solitary; sepals ovate, c. 0.5 cm long, purple with green apices; petals oblong-ovate, c. 1.5 mm long, purple; lip tongue-shaped to oblong, c. 2.5–3 mm long, papillose, fleshy, purple with a yellow apex; column 2 mm long. (See fig. 27).

Fig. 27. *Bulbophyllum betchei*. **A**, habit × ⅔; **B**, flower × 6; **C**, lip × 10; **D**, column × 14; **E**, anther cap × 14; **F**, pollinia × 14; **G**, dorsal sepal × 6; **H**, lateral sepal × 6; **J**, petal × 6; **K**, column and lip × 10. *B. samoanum*. **L**, habit × ⅔; **M**, lip × 6; **N**, column and lip × 6; **O**, flower × 3; **P**, column × 6, **Q**, anther cap × 10; **R**, pollinia × 10; **S**, dorsal sepal × 4; **T**, petal × 4; **U**, lateral sepal × 4. **A** drawn from *MacKee* 19; **B–K** from *Sprunger* 206 (Kew spirit no. 51927); **L** from *Whistler* 2787; **M–U** from *Sprunger* 121 (Kew spirit no. 51889). All drawn by Sue Wickison.

DISTRIBUTION: Anatom, Efate and Espiritu Santo. Also in Caroline Islands, Solomon Islands, New Caledonia and Samoa.
HABITAT: Montane forest, 460–700 m.
COLLECTIONS: *Cribb* S4 (K); *Cribb and Wheatley* 29 & 68 (K); *Hoock* s.n. (P).

This species is similar to *Bulbophyllum radicans* Bailey from Australia, but it differs in having shorter papillae on the lip.

section PELMA

11. B. neocaledonicum *Schltr.* in Engler, Bot. Jahrb. 39: 84 (1906). Type: New Caledonia, *Schlechter* 15492 (holotype B; isotypes K!, P!, Z).
Pelma neocaledonica (Schltr.) Finet in Not. Syst. 1: 113; fig. 6: 114 (1909).

Rhizome of 2 types, creeping and leafless below and erect, leaf-bearing, branched and covered in cataphylls above. *Pseudobulbs* c. 3 mm high, 2 mm wide, c. 3 mm apart. *Leaf* lanceolate-oblong, c. 17 mm long, 3 mm wide, olive-green. *Inflorescence* arising from rhizome, c. 1 cm long. *Flower* apical, solitary, white to white-green; sepals ovate, c. 2–3 mm long; petals ovate, c. 1–1.5 mm long; lip oblong, c. 7 mm long with 2 raised calli at the base; column c. 4 mm long.

DISTRIBUTION: Espiritu Santo. Also in New Caledonia.
HABITAT: Rain forest, 1050 m.
COLLECTION: *Cabalion* 2766 (K, P).

12. B. savaiense *Schltr.* in Fedde, Rep. Sp. Nov. 9: 106 (1911). Type: Samoa, *Vaupel* 596 (holotype B).

Rhizome erect, bearing many cataphylls and adventitious roots, 2–5 cm high. *Pseudobulbs* ovoid, c. 4 mm high, 2 mm wide, appressed to rhizome, clustered. *Leaf* ovate, 1–2 cm long, 4–5 mm wide. *Inflorescence* arising from the rhizome, 3–7 cm long. *Flowers* 4–6, white, miniscule; sepals ovate, 1–2 mm long; petals ovate, c. 1 mm long; lip circular-ovate with a concavity at the base in which column-foot fits, c. 1 mm long; column c. 1 mm long.

DISTRIBUTION: Erromango and Pentecost. Also in Fiji and Samoa.
HABITAT: Rain forest, c. 330 m.
COLLECTIONS: *Raynal* in RSNH 16239 (K, P); *Wheatley* 248 (K, PVNH).

This species is similar to *Bulbophyllum foveatum* Schltr. from New Guinea.

section MACROUSIS

13. B. pensile *Schltr.* in Fedde, Rep. Sp. Nov., Beih. 1: 866 (1913). Type: New Guinea, *Schlechter* 19895 (holotype B).
Bulbophyllum levatii Kraenzl. in Bull. Soc. Bot. Fr. 76: 300 (1929); **synon. nov.** Type: Vanuatu, *Levat* s.n. (holotype MPU).

Rhizome 1.5–2 mm wide, woody. *Pseudobulbs* ovate, c. 7 mm high, 7 mm wide, yellow-green to green, 2–4 cm apart. *Leaf* oblong, 3–7 cm long, 0.8–1.2 cm wide; petiole c. 1 cm long. *Inflorescence* from the base of pseudobulb; peduncle 10–12 cm long; rhachis purple basally, paler to white apically. *Flowers* 4–7, white; sepals lanceolate, acuminate, c. 1.2 cm long; petals ovate, c. 3 mm long; lip tongue-shaped with 2 ridged keels, c. 2–3 mm long; column c. 2 mm long.

DISTRIBUTION: Ambae, Efate, Erromango, Espiritu Santo and Pentecost. Also in New Guinea and the Solomon Islands.

HABITAT: Rain forest, 200–1400 m.

COLLECTIONS: *Bregulla* 30 (K); *Cabalion* 1097 (NOU); *Cribb & A. Morrison* 1821 (K); *Cribb & Wheatley* 27 & 113 (K, PVNH); *Green* in RSNH 1057,1177 & 1353 (K); *Morat* 5435 (P); *Raynal* in RSNH 15998 (P); *Suprin* 301 (P); *Wheatley* 83 & 301 (K, PVNH).

This species may be conspecific with *Bulbophyllum cavistigma* J.J. Smith from New Guinea.

14. B. setipes *Schltr.* in Fedde, Rep. Sp. Nov., Beih. 1: 869 (1913). Type: New Guinea, *Schlechter* 18684 (holotype B).

Rhizome 1.5–2.0 mm wide, woody. *Pseudobulbs* ovate, c. 0.5 cm high, 0.4 cm wide, 2–3.5 cm apart. *Leaf* oblong-lanceolate, (1.5)2–6 cm long, 0.6–1.3 cm wide. *Inflorescence* from base of pseudobulb; peduncle 10–18 cm long; rhachis zig-zag when in flower, straight in fruit. *Flowers* 4–6, yellow with maroon stripes; sepals ovate , c. 4 mm long; petals oblong c. 2mm long; lip tongue-shaped with 2 ridged keels, 2–3 mm long, greenish; column c. 1.5 mm long.

DISTRIBUTION: Banks Islands (Vanua Lava), Erromango, Espiritu Santo and Pentecost. Also in New Guinea and the Solomon Islands.

HABITAT: *Agathis, Calophyllum neoebudicum* community and coconut plantations, sea level to 900 m.

COLLECTIONS: *Bernardi* 13239 (G, P); *Cribb & Wheatley* 61 (K, PVNH); *Hoock* s.n. (P); *McKee* in RSNH 24162 (P); *Wheatley* 361 (K, PVNH).

section APHANOBULBON

15. B. polypodioides *Schltr.* in Engler, Bot. Jahrb. 39: 86 (1906). Type: New Caledonia, *Schlechter* 15422 (holotype B, isotypes K!, P!, Z).
B. nigroscapum Ames in Orchid. 7: 86 (1922). Type: Samoa, *Collarino* 383 (holotype AMES).

Rhizome bearing many adventitious roots. *Pseudobulbs* very reduced. *Leaf* oblong, 5–12.5 cm long, 1.5–2.5 cm wide, pale green; petiole 2–5 cm long. *Inflorescence* from base of pseudobulb; peduncle may be covered in bracts, 5–11 cm long (often black in dried specimens); rhachis, 5–10 cm long, pale green. *Flowers* 10–20, yellow to pale green; sepals lanceolate, c. 4 mm long; petals lanceolate-oblong, 2.4–3 mm long; mentum c. 1.5 mm long; lip tongue-shaped, 2.5–3 mm long, bright yellow; column c. 1 mm long.

DISTRIBUTION: Ambae, Espiritu Santo and Pentecost. Also in the Solomon Islands, New Caledonia, Fiji and Samoa.

HABITAT: Ridge-top forest, 500–800 m.

COLLECTIONS: *Cribb & Wheatley* 59 & 118 (K, PVNH); *Morat* 5223 (P); *Wheatley* 99 (K, PVNH).

This species is similar to *Bulbophyllum tahitiense* Nadeaud, from Tahiti.

section GLOBICEPS

16. B. atrorubens *Schltr.* in Engler, Bot. Jahrb. 39: 82 (1906). Type: New Caledonia, *Schlechter* 15495 (holotype B, isotypes K!, P!).

Rhizome bearing many adventitious roots. *Pseudobulbs* 0.5 cm high, 0.3 cm wide, clustered. *Leaf* obovate, 4–6 cm long, 1.7–2.3 cm wide, succulent, yellow-green; petiole c. 2 cm long. *Inflorescence* from the base of the pseudobulb; peduncle 10–20 cm long. *Flowers* apical in a dense globose head, maroon; sepals ovate, c. 3.5 mm long; petals spathulate, c. 1.5 cm long; lip circular, arched in profile, 1.5 mm long, fleshy, papillose; column c. 1 mm long; anther-cap pale yellow.

DISTRIBUTION: Espiritu Santo. Also in New Caledonia.
HABITAT: Montane ridge-top forest, c. 700 m.
COLLECTIONS: *Cribb & Wheatley* 32 & 33 (K, PVNH).

section CIRRHOPETALUM

17. B. longiflorum *Thouars*, Orch. Iles Austr. Afr.: 98 (1822). Type: Mauritius, *Thouars* s.n. (holotype P).
Epidendrum umbellatum G. Forst., Ins. Austr. Prodr.: 60 (1786). Type: Society Islands, *Forster* s.n. (holotype BM!).
Cymbidium umbellatum (G. Forst.) Spreng., in Pl. Min. Cog. Pugill. 2: 82 (1815).
Cirrhopetalum thouarsii Lindley in Bot. Reg 10: sub. t. 832 (1824). Type: Fiji, *Seemann* 598 (holotype K!).
Cirrhopetalum umbellatum (G. Forst.) Hooker & Arn., Bot. Beechey Voy.: 71 (1832).
Cirrhopetalum clavigerum Fitzg. in Journ. Bot. 21: 204 (1883). Type: N. Australia, *Fitzgerald* s.n. (holotype BM).
Bulbophyllum clavigerum (Fitzg.) F. Muell. in Syst. Cens. Austr. Pl. Suppl. 1: 3 (1884).
Phyllorchis umbellata (G. Forst.) Kuntze, Rev. Gen. Pl. 2: 657 (1891).
Phyllorchis thouarsii (Lindley) Kuntze, l.c. 677.
Phyllorchis clavigera (Fitzg.) Kuntze, l.c.
Cirrhopetalum thouarsii var. *concolor* Rolfe in Gard. Chron. ser. 3, 12: 178 (1892). Type: not found.
Cirrhopetalum kenejianum Schltr. in Fedde, Rep. Sp. Nov., Beih. 1: 889 (1913). Type: New Guinea, *Schlechter* 18462 (holotype B).
Cirrhopetalum longiflorum (Thou.) Schltr. in Beih. Bot. Centralbl. 33, 2: 420 (1915).

Rhizome thick and woody, 5–9 mm diameter. *Pseudobulbs* conical, 1.5–3.5 cm high, 0.8–1.2 cm wide, often covered with hair-like remains of sheathing bracts, usually 5–6 cm apart but may be closer. *Leaf* oblong, 9–16 cm long, 2–3.5 cm wide, thick; petiole 1.5–3 cm long. *Inflorescence* arising from base of pseudobulb; umbellate; peduncle up to 30 cm long. *Flowers* 6–8, cream-yellow, blotched and speckled deep maroon-red; dorsal sepal ovate, concave with an apical cilia, c. 8 mm long; lateral sepals linear, fused, 2.5–3.5 cm long; petals ovate, c. 6–9 mm long, purple, with ciliate margins; lip tongue-shaped, c. 4–5 mm long, purple; column c. 3 mm long, with c. 4 mm long, slightly downcurved stelidia.

DISTRIBUTION: Efate. Widesread from Africa and Madagascar to the Malay archipelago, New Guinea, Fiji and Australia.
HABITAT: Montane forest growing in plenty of light, 200 m.

COLLECTIONS: *Bregulla* 15 (K); *MacKee* 32755 (P); *Raynal* in RSNH 2311 & 17962 (P).

53. **PEDILOCHILUS** Schlechter

A genus closely related to *Bulbophyllum* but distinguished by possessing a saccate lip, the petals are twisted slightly into the shape of the letter S.

A genus of about 15 species from New Guinea and the adjacent islands. A new genus record for Vanuatu, a single species being recorded.

P. hermonii *Cribb & B. Lewis* in Orchid Rev. in press (1989). Type: Vanuatu, Espiritu Santo, *Cribb & Wheatley* 14 (holotype K!).

A small *epiphyte* with a short creeping rhizome and closely spaced erect ovoid or obliquely ovoid pseudobulbs, 1–1.5 cm tall, 0.4–0.7 cm wide, 1-leaved at apex and sheathed by papery bracts at the base. *Leaf* erect, coriaceous, linear to narrowly oblong-elliptic, obtuse, 3–12 cm long, 1–1.5 cm wide; petiole 0.2–1.5 cm long. *Inflorescence* 1-flowered, basal, 2–6 cm long; bract obconical, obtuse, 5 mm long. *Flower* whitish on outer surface, but purple or purple-veined inside the sepals and petals and with a purple lip; pedicel and ovary 4–8 mm long; dorsal sepal elliptic to obovate, obtuse, 8–10 mm long, 4–5 mm wide, ciliolate; lateral sepals obliquely oblong-elliptic, obtuse, 9–11 mm long, 5.5–6.5 mm wide, obscurely ciliolate; petals shortly clawed, ovate, subacute, 6–7 mm long, 3 mm wide; lip clawed, urceolate like a small slipper, obtuse at reflexed apex, 8–10 mm long, 6 mm wide, with recurved margins, with small recurved auricles at either side at the apex of the claw and a low fleshy callus between them; column 2 mm long, with long attenuate apical stelidia, 1.5 mm long. (See fig. 28, plate 7d).

DISTRIBUTION: Anatom and Espiritu Santo.
HABITAT: Montane, ridge-top and summit cloud forest, 700–1800 m.
COLLECTIONS: *Cabalion* 452 (PVNH); *Cribb & Wheatley* 14 & 88 (K, PVNH); *Morrison* in RBG Kew 40 (K); *Raynal* in RSNH 16336 & 16337 (K, P, PVNH); *Veillon* 2457 (P) & 4076 (NOU, P).

Pedilochilus hermonii is closest to the New Guinea species *P. ciliolatum* Schltr. sharing the minutely ciliate sepals and tubular obconical bract of that species. However, *P. hermonii* has a shorter inflorescence, a more deeply saccate, broader lip, longer recurved basal auricles on the claw of the lip and a shorter broader recurved apex. Its flowers are also distinctively coloured, plum-purple within but whitish or very pale purplish on the outside.

54. **GUNNARELLA** Senghas

Epiphytic. Stem monopodial, short. *Inflorescences* lateral, pendent. *Flowers* many, not opening widely, diaphanous; sepals and petals free, about equal in length; lip trilobed, hinged to the column-foot, lateral lobes erect; column short with a

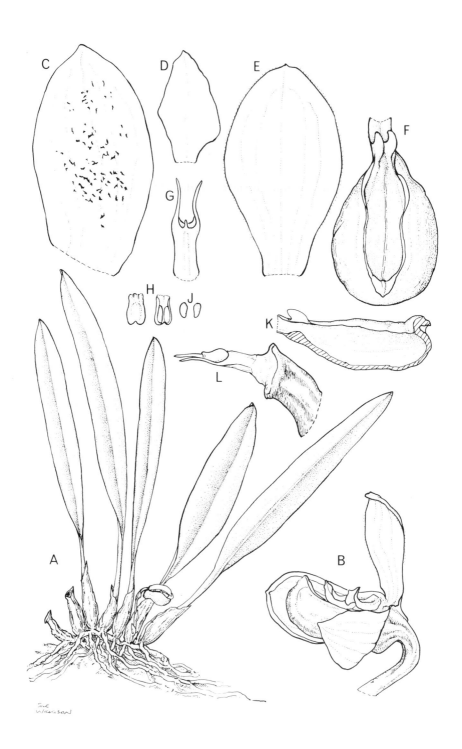

Sue
wilkinson

136

short foot; pollinia 4.

A small genus of about 6 species in New Guinea and the Pacific Islands. A single species in Vanuatu.

G. robertsii (*Schltr.*) *Senghas* in Die Orchidee 39(2): 73 (1988). Type: New Caledonia, *Roberts* s.n. (holotype B).
Sarcochilus robertsii Schltr. in Fedde, Rep. Sp. Nov. 3: 320 (1907).
Chamaeanthus robertsii (Schltr.) Schltr. in Fedde, Rep. Sp. Nov., Beih 1: 957 (1913).
Chamaeanthus laxus Schltr., l.c. 958. Type: New Guinea, *Schlechter* 18367 (holotype B).
Gunnarella laxus (Schltr.) Senghas in Die Orchidee 39(2): 73 (1988).

Stem 1–2 cm long. *Leaves* up to 10, arranged in a fan, linear, 6–13 cm long, 0.7–1 cm wide, obliquely bilobed at apex, green or reddish. *Inflorescence* up to 28 cm long, pendent. *Flowers* 9–25, translucent white; sepals and petals linear-lanceolate, c. 4 mm long; lip ovate, obtuse c. 1.8 mm long, 0.8 mm wide, fleshy at base, lateral lobes erect. (See fig. 29, plate 7b).

DISTRIBUTION: Efate, Erromango and Espiritu Santo. Also in New Guinea, the Solomon Islands and New Caledonia.
HABITAT: Rain forest near river, c. 150 m.
COLLECTIONS: *Bourdy* 190 (P); *Cribb & Wheatley* 5 (K, PVNH); *Hallé* in RSNH 6465 (P).

55. **SARCANTHOPSIS** Garay

Erect, stout *epiphytic* plants. *Leaves* broad, strap-shaped; internodes short. *Inflorescences* lateral, branched, stout, with many flowers. *Flowers* fleshy; sepals and petals almost equal and widely spreading; lip attached to the base of the column by the lateral lobes, midlobe a fleshy flap which recurves over the basal keel of the lip; column short; pollinia 4.

A small genus of about 5 species from New Guinea, the Moluccas and the Pacific Islands. A single species in Vanuatu.

S. nagarensis (*Reichb. f.*) *Garay* in Bot. Mus. Leafl. Harv. Univ. 23(4): 199 (1972). Type: Fiji, *Seemann* 594 (holotype K!).
Sarcanthus nagarensis Reichb. f. in Seem., Fl. Vit.: 298 (1868). Type: Fiji, *Seemann* 594 (holotype K!).
Stauropsis warocqueana Rolfe in Lindenia 7: 65, t.319 (1891). Type: New Guinea, cult. l'Hort. Internationale (holotype K!).
Vandopsis warocqueana (Rolfe) Schltr. in K. Schum. & Laut., Nachtr. Fl. Deutsch. Sudsee: 225 (1905).
Stauropsis woodfordii Rolfe in Kew Bull. 1908: 72 (1908). Type: Solomon Islands, *Officers of H.M.S. Penguin* s.n. (holotype K!).
Stauropsis nagarensis (Reichb. f) Rolfe in Kew Bull. 1909: 64 (1909).

Fig. 28. *Pedilochilus hermonii*. **A**, habit × ⅔; **B**, flower × 3; **C**, lateral sepal × 4; **D**, petal × 4; **E**, dorsal sepal × 4; **F**, lip × 4; **G**, column × 6; **H**, anther cap × 6; **J**, pollinia × 6; **K**, cross section lip × 4; **L**, column × 6. Drawn from *Cribb & Wheatley* 14 (Kew spirit no. 53174) by Sue Wickison.

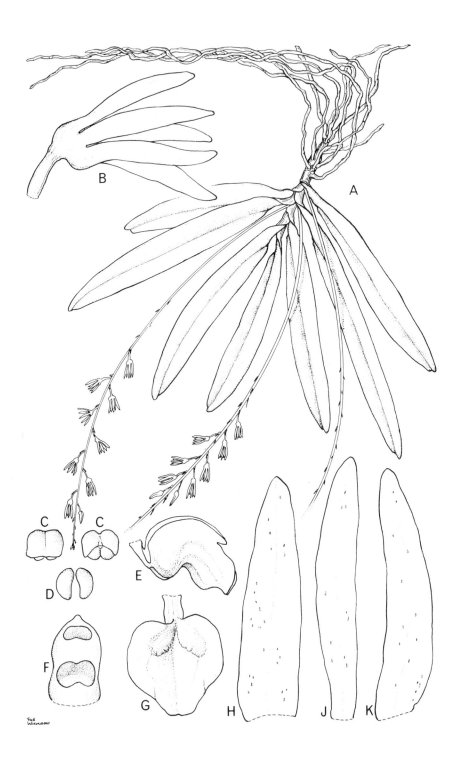

Stauropsis quaifei Rolfe in Kew Bull. 1909: 64 (1909). Type: Vanuatu, Espiritu Santo, *Quaife* 265 (holotype K!).
Vandopsis quaifei (Rolfe) Schltr. in Fedde, Rep. Sp. Nov., Beih. 10: 196 (1911).
Vandopsis nagarensis (Rolfe) Schltr. in Fedde, Rep. Sp. Nov., Beih. 1: 972 (1913).
Vandopsis raymundi Schltr. in Engler, Bot. Jahrb. 58: 494 (1921). Type: Caroline Islands, *Ledermann* s.n. (holotype K!).
Sarcanthopsis quaifei (Rolfe) Garay in Bot. Mus. Leafl. Harv. Univ. 23(4): 199 (1972).
Sarcanthopsis warocqueana (Rolfe) Garay, l.c.
Sarcanthopsis woodfordii (Rolfe) Garay, l.c.

Epiphytic. Stems up to 3.5 m long, 2.5 cm wide. *Leaves* alternate, strap-like, up to 40 cm long, 6 cm wide, fleshy, dark green, rounded at apex; internodes c. 5 cm long. *Inflorescence* emerging from stem near apex, opposite leaf axil, up to 40 cm long, branched. *Flowers* in clusters at apex of branches, c. 30, yellow spotted with reddish-brown, with lip and column rather paler with purplish markings and lines, scented; pedicels c. 1.5 cm, ridged; sepals and petals oblong-spathulate, c. 8 mm long; lip c. 8 mm long when flattened, lateral lobes erect, midlobe convex, fleshy. (See fig. 30).

DISTRIBUTION: Anatom, Efate, Epi, Erromango, Espiritu Santo, Pentecost and Tanna. Also in New Guinea, Bougainville, the Solomon Islands, the Caroline Islands, the Horn Islands and Fiji.
HABITAT: Open localities, exposed to direct sun (for example on the side of roads), sea level to 1370 m.
COLLECTIONS: *Bernardi* 13082 (K, G, P); *Bregulla* 28 (K); *McKee* 41298 (P); *Morrison* in RBG Kew 79 (K); *Quaife* 265 (K); *Im Thurn* 352 (K); *de la Rüe* s.n. (P); *Suprin* 378 (P); *Wheatley* 206 & 262 (K, PVNH).

G. Dennis (pers. comm.) notes that the flowers tend to face the light.

56. **THRIXSPERMUM** Loureiro

Epiphytic. Stem short or long, with a few leaves close together or distichously arranged along the stem. *Inflorescence* lateral. *Flowers* close together or distichous, with persistent bracts; sepals and petals free, more or less equal; lip immovably joined to the column-foot, saccate; column short with a distinct broad foot; pollinia 4.
A genus of about 100 species from tropical Asia to Australia and the Pacific Islands. Two species are recorded from Vanuatu, both being new records. The species are arranged in 2 sections according to Schlechter (1913).

Flowers and bracts in 2 ranks   ... ... **1. T. graeffei**
Flowers and bracts spirally arranged   ... ... **2. T. adenotrichium**

Fig. 29. *Gunnarella robertsii*. **A**, habit × ⅔; **B**, flower × 6; **C**, anther cap × 18; **D**, pollinia × 18; **E**, lip side view × 14; **F**, column × 18; **G**, lip × 14; **H**, dorsal sepal × 12; **J**, petal × 12; **K**, lateral sepal × 12. Drawn from *Cribb & Wheatley* 5 (Kew spirit no. 53175) by Sue Wickison.

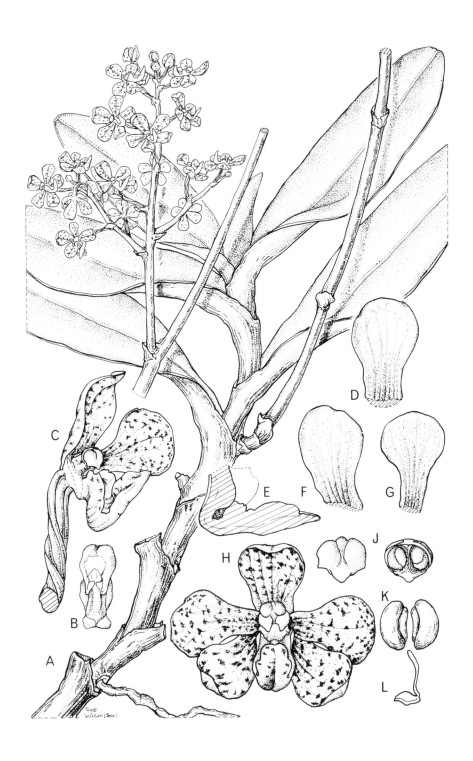

section ORSIDICE

1. T. graeffei *Reichb. f.* in Seem., Fl. Vit.: 298 (1868). Type: Samoa, *Graeffe* s.n. (holotype W).
Sarcochilus graeffei (Reichb. f.) Benth. & Hook. f. ex Drake, Ill. Ins. Mar. Pacif.: 310 (1892).
Thrixspermum oreadum Schltr. in Fedde, Rep. Sp. Nov., Beih. 1: 962 (1913). Type: New Guinea, *Schlechter* 16927 (holotype B).

Stem up to 8 cm high. *Leaves* 3–5, 3–6 cm long, c. 0.8 cm wide. *Inflorescences* up to 9 per stem; peduncle up to 10 cm long; raceme bilaterally flattened, up to 5 cm long; bracts alternate. *Flowers* in 2 alternating rows, pale yellow; lip creamy-white with orange-ochre blotches; sepals and petals concave and cucullate, ovate, c. 5.5 mm long; lip oblong, saccate, c. 8 mm long, 4.5 mm wide, with incurved margins, with a basal callus, and 2 fleshy lateral ridges midway down lip, papillose at apex. *Fruits* up to 4.5 cm long, 3 mm wide.

DISTRIBUTION: Efate and Erromango. Also in New Guinea, the Solomon Islands, Samoa and Fiji.
HABITAT: Forest and hills.
COLLECTIONS: *Cribb & A. Morrison* 1803a (K); *Morrison* in RBG Kew 28, 30 & 33 (K); *Raynal* in RSNH 16226 (K, P).

section DENDROCOLLA

2. T. adenotrichum *Schltr.* in Fedde, Rep. Sp. Nov., Beih. 1: 962 (1913). Type: New Guinea, *Schlechter* 18923 (holotype B).

Stem up to 1 cm long. *Leaves* up to 2.5 cm long, 3 mm wide. *Inflorescence* c. 3 cm long; raceme c. 5 mm long; bracts spirally arranged. *Flowers* white to pale yellow-orange; lip may be white with brown dots, or margins may be pale pink; sepals and petals ovate, 6–7 mm long; lip trilobed, hairy, saccate, 6–7 mm long, 9–10 mm wide, lateral lobes obliquely oblong, obtuse, midlobe oblong, obtuse, with a basal callus. *Fruit* 2.5 cm long, 1.5 mm wide.

DISTRIBUTION: Erromango. Also in New Guinea, the Solomon Islands and New Caledonia.
HABITAT: *Agathis* forest.
COLLECTION: *Bernardi* 13237 (G, P) (in fruit).

The description of the flowers is based on Solomon Island specimens.
This species corresponds with *Thrixspermum sp.*, of Hallé (1977). It may also be conspecific with *T. congestum* (Bail.) Dockr., from Australia.

Fig. 30. *Sarcanthopsis nagarensis*. **A**, habit ×⅔; **B**, lip from above × 3; **C**, flower with lateral sepal and petal removed × 3; **D**, dorsal sepal × 3; **E**, longitudinal section lip × 4; **F**, lateral sepal × 3; **G**, petal × 3; **H**, flower × 3; **J**, anther cap × 6; **K**, pollinia × 10; **L**, stipe × 10. All drawn from *Wickison & Morrow* 140 (Kew spirit no. 51498) by Sue Wickison.

57. **SCHOENORCHIS** Reinwardt

Epiphytic. Stems erect or pendulous, often branched near the base. *Leaves* distichous, linear to terete. *Inflorescences* lateral, multiflowered, racemose; peduncle shorter than rhachis. *Flowers* minute; sepals and petals free; lip spurred; column short, footless; pollinia 4.

A genus of about 10 species from S.E. Asia to Australia and the Pacific Islands. A single species in Vanuatu.

S. micrantha *Reinw. ex Blume* in Bijdr. 1(8): 362 (1825). Type: Java, *Blume* s.n. (holotype L).
Saccolabium chionanthum Lindley in Journ. Linn. Soc. 3: 35 (1859). Type: Java, *Reinwardt* s.n. (drawing K!).
Schoenorchis densiflora Schltr. in Fedde, Rep. Sp. Nov., Beih. 1: 986 (1912). Type: New Guinea, *Schlechter* 17316 (isotype K!).
Saccolabium plebejum J.J. Smith in Bull. Jard. Bot. Buitenz. ser. 2, 3: 77 (1912). Type: New Guinea, *Gjellerup* 21 (holotype BO).
Ascocentrum micranthum (Reinw. ex Blume) Holtt. in Gard. Bull. Sing. 11: 275 (1947).

Plant growing in small tangled clumps, due to progressive branching of the stem. *Stem* 2–15 cm long, leafy throughout. *Leaves* linear, very thick, sulcate, grooved above, 3–6 cm long, 1 mm wide. *Inflorescences* 1–4, 1–2.5 cm long. *Flowers* 6–30, white, bell-shaped; dorsal sepal lanceolate, c. 2 mm long; lateral sepals oblong, c. 2 mm long; petals not quite as long as the sepals but a little broader; lip trilobed, 2.5 mm long, 1.5 mm wide, with low, erect lateral lobes, longer than the midlobe, midlobe acute and recurved at apex, spur globose to ellipsoid.

DISTRIBUTION: Erromango. Widely distributed from Asia, S.E. Asia and the Malay archipelago to New Guinea, Bougainville, the Solomon Islands, New Caledonia, Fiji and Australia.
HABITAT: Rain forest.
COLLECTION: *Bernardi* 13194 (K, G, P).

58. **LUISIA** Gaudichaud

Epiphytic. Stems long. *Leaves* terete, distichous, leaf bases sheathing. *Inflorescences* lateral, very short, with flowers usually developing one or a few at a time. *Flowers* small, opening widely; sepals and petals free; petals often longer and narrower than the sepals; lip fleshy, deflexed, divided into 2 distinct portions, basal and apical, the basal portion narrower than the apical, concave, with distinct lateral lobes, apical portion expanded into a somewhat broad lamina which is thicker than the basal portion; column short with a short foot; pollinia 4.

A genus of about 40 species from tropical Asia to the Pacific Islands. A single species in Vanuatu.

L. teretifolia *Gaud.* in Freyc., Voy. Bot.: 427, t.37 (1826). Type: Mariana Islands, Guam, *Gaudichaud* 37 (holotype P!, sterile).

Stem erect, woody, 10–30 cm long. *Leaves* distichous, terete, rigid, 5–15 cm long, 2–5 mm wide, dark green. *Inflorescences* 5–15 mm long. *Flowers* open 1–3 at a time;

sepals and petals yellow to green (may have purple mottling inside), with apex of lip purple or rarely green; pedicels 6–8 mm long; sepals 5–7 mm long; petals 8–9 mm long; lip 6–7 mm long, 6–7 mm wide; column 4 mm long, foot c. 2 mm long.

DISTRIBUTION: Efate, Espiritu Santo and Pentecost. Widely distributed from S.E. Asia to the Malay archipelago, New Guinea, the Solomon Islands, the Mariana Islands (Guam), New Caledonia and Australia (see note).

HABITAT: Near coast in open situations or in open woodland of *Casuarina* by river, sea level to 310 m.

COLLECTIONS: *Bregulla* 27 (K); *Cribb & Morrison* 1779 (K); *Cribb & Wheatley* 123 & 125 (K, PVNH); *Im Thurn* 337 (K); *Suprin* 273 (P); *Wheatley* 192 (K, PVNH).

Seidenfaden (1988) notes that a large number of small-flowered *Luisia* specimens have through the years been recorded under this name, resulting in a large distribution area for the species. This is probably incorrect, and it is uncertain how many specimens can rightly be referred to *L. teretifolia*.

59. **CLEISOSTOMA** Blume

Epiphytic. Stems stout. *Leaves* oblong. *Inflorescence* lateral, usually branched. *Flowers* with sepals and petals free; lip with a rounded saccate spur; the front wall of the spur has an erect tongue; column short; pollinia 4.

A genus of 80–100 species from tropical Asia to the Pacific Islands. A new genus record for Vanuatu, a single species being recorded.

C. pacificum *Cribb & B. Lewis* in Orchid Rev. in press (1989). Type: Vanuatu, Pentecost, *Wheatley* 302 (holotype K!).

Stem stout, erect or pendent-arcuate, leafy, 10–30 cm long, 0.7–1.4 cm wide, covered by sheathing leaf bases. *Leaves* distichous, spreading, coriaceous, linear, unequally roundly bilobed at the apex, 18–33 cm long, 1.5–2.2 cm wide. *Inflorescence* axillary, emerging through the leaf sheaths opposite the leaves, paniculate or rarely unbranched; peduncle terete, 17–25 cm long; branches 2.5–7 cm long; bracts triangular, spreading or reflexed, 1–2 mm long. *Flowers* small, with whitish, pale green or yellow sepals and petals and a white lip; pedicel and ovary 4–5 mm long; dorsal sepal elliptic, obtuse or rounded, 3–4 mm long, 2 mm wide; lateral sepals obliquely oblong-elliptic, obtuse, 4–5 mm long, 2–2.5 mm wide; petals oblong, obtuse, 3 mm long, 1 mm wide; lip very fleshy, united on basal margins to the column, ovate, obtuse, 2–2.5 mm long and wide, spur nectariferous, shortly conical-cylindrical, obtuse, 3.5–4 mm long, with a high transverse callus at the base of the lip almost blocking the mouth of the spur; column fleshy, 1–1.3 mm long. (See fig. 31).

DISTRIBUTION: Pentecost. Also in the Solomon Islands, New Caledonia and Fiji.

HABITAT: Rain forest, epiphytic on *Gyrocarpus americanus*, near the base, 350 m.

COLLECTIONS: *Wheatley* 253 & 302 (K, PVNH).

Cleisostoma pacificum is closely allied to the Javanese species *C. montanum* (J.J. Smith) Garay but differs in having flowers in which the spur is not globose and

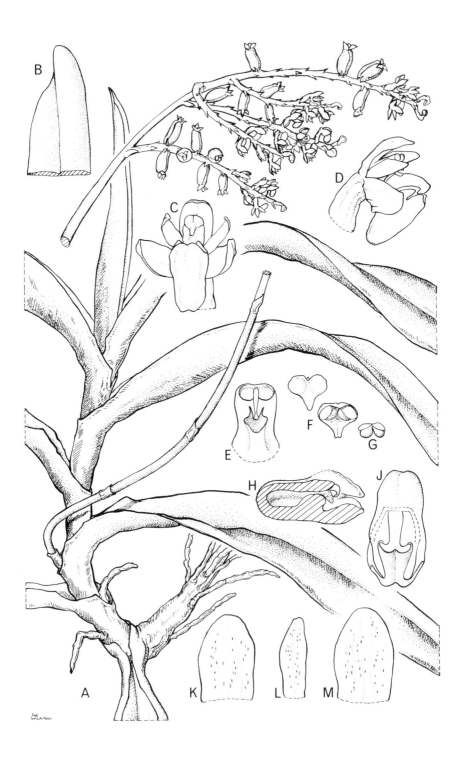

lacks a callus on the upper side of the mouth of the spur but has a higher transverse fleshy callus at the base of the lip that almost blocks the mouth of the spur. The spur in *C. pacificum* lies parallel to the ovary and not at right angles to it as in *C. montanum.*

60. **MICROTATORCHIS** Schlechter

Small *epiphytic* herbs. *Stems* short. *Leaves* on young plants only. *Raceme* lateral, angled; bracts persistent, alternate, distichous. *Flowers* not opening widely; sepals and petals fused at base; lip shortly spurred; column short; pollinia 4.

A genus of about 25 species from New Guinea to New Caledonia. A new genus record for Vanuatu, a single species being recorded.

M. schlechteri *Garay* in Bot. Mus. Leafl. Harv. Univ. 23(4): 187 (1972). Types: as for *M. fasciola* Schltr.

Microtatorchis fasciola Schltr. in Engler, Bot. Jahrb. 39: 88 (1906). Types: New Caledonia, *Schlechter* 14911, 15300 & 15665 (isosyntypes BM, K!, P, Z) non (G. Forst.) Schltr. (1905).

Stem up to 1 cm long. *Roots* may be green. *Leaves* 2 cm long, 0.4 cm wide, light green. *Racemes* up to 6, up to 10 cm long, may appear zig-zag; bracts at a 45–90 degree angle to the raceme. *Flowers* c. 4, pale yellow to green; sepals lanceolate, c. 3 mm long; petals lanceolate, c. 2.5 mm long; lip triangular, acuminate, c. 3 mm long, 1 mm wide, margins inrolled, spur globose. *Fruits* up to 1.5 cm long, 2 mm wide. (See fig. 32, plate 8c).

DISTRIBUTION: Ambae, Espiritu Santo and Pentecost. Also in New Caledonia and Fiji.

HABITAT: Montane forest, 1300–1500 m.

COLLECTION: *Cabalion* 2609 (P); *Cribb & Wheatley* 76 (K, PVNH); *Wheatley* 60 (K, PVNH).

61. **POMATOCALPA** Breda

Epiphytic. Stems stout. *Leaves* oblong. *Inflorescence* branched, densely many-flowered. *Flowers* small; sepals and petals free and almost spreading; lip trilobed with a rounded, saccate spur, lateral lobes small, joined at the back to the base of the column, the front edges incurved, the back wall of the spur has an erect tongue, which reaches to the mouth of the spur; column short; pollinia 4.

A genus of about 30 species from S.E. Asia to Australia and the Pacific Islands. A single species in Vanuatu.

Fig. 31. *Cleisostoma pacificum.* **A,** habit × ⅔; **B,** apex of leaf × ⅔; **C,** flower × 4; **D,** flower side view × 4; **E,** column × 8; **F,** anther cap × 8; **G,** pollinia × 8; **H,** cross section lip × 6; **J,** lip × 6; **K,** dorsal sepal × 6; **L,** petal × 6; **M,** lateral sepal × 6. Drawn from *Wickison* 49 (Kew spirit no. 50906) by Sue Wickison.

P. marsupiale (*Kraenzl.*) *J.J. Smith* in Herderschee, Nova Guinea 7: 101 (1913). *Cleisostoma marsupiale* Kraenzl. in K. Schum. & Hollr., Fl. Kais. Wilh. Land: 34 (1889).Type: New Guinea, *Hollrung* 743 (holotype B).

Inflorescence branched, 30 cm long, branches 7–8 cm long. *Flowers* numerous, 60–80, dense, maturing at base first; sepals and petals green, with a white lip; dorsal sepal oblong, c. 5 mm long; petals oblong-lanceolate, 3 mm long; lateral sepals oblong, 5 mm long; lip c. 3 mm long, 3 mm wide, spurred.

DISTRIBUTION: Espiritu Santo. Also in New Guinea and the Solomon Islands.
HABITAT: Open localities, lowland to moderate altitudes.
COLLECTION: *Bregulla* 25 (K) (inflorescence only).

We have not seen the habit of this collection and the flowers are rather smaller than those of Solomon Island collections of *Pomatocalpa marsupiale*. More complete material is needed to confirm the identification of the Vanuatu taxon.

62. **ROBIQUETIA** Gaudichaud

Epiphytic. Stems stout, pendulous. *Leaves* broadly oblong. *Inflorescence* erect or pendulous, unbranched, with many crowded small flowers. *Flowers* non-resupinate, with a large spur uppermost in flower; column short without a foot; pollinia 4.
A genus of about 25 species from S.E. Asia to Australia and the Pacific Islands. A single species in Vanuatu.

R. bertholdii (*Reichb. f.*) *Schltr.* in Fedde, Rep. Sp. Nov., Beih. 1: 983 (1913). *Saccolabium bertholdii* Reichb. f. in Seem., Fl. Vit.: 595 (1868). Type: Fiji, *Seemann* 595 (holotype K!). *Saccolabium constrictum* Reichb. f. in Otia Bot. Hamb. 1: 52 (1878). Type: Fiji, *C. Wilkes* s.n. (holotype W). *Saccolabium mimus* Reichb. f. in Gard. Chron. n.s. 9: 266 (1878). Type: Pacific Islands, *P. Veitch* s.n (holotype W). *Saccolabium graeffei* Reichb. f. in Gard. Chron. n.s. 16: 716 (1881). Type: Fiji, *Graeffe* s.n. (holotype W). *Malleola mimus* (Reichb. f.) P.F. Hunt in Kew Bull. 24: 98 (1970). *Saccolabium kajewskii* Ames in Journ. Arn. Arb. 13: 141 (1932). Type: Vanuatu, Efate, *Kajewski* 205 (holotype AMES).

Fig. 32. *Microtatorchis schlechteri.* **A**, habit life size; **B**, flower × 8; **C**, lip and spur × 14; **D**, column × 14; **E**, column side view × 14; **F**, anther cap × 22; **G**, pollinia × 22; **H**, stipe × 22; **J**, dorsal sepal × 14; **K**, petal × 14; **L**, lateral sepal × 14. *Taeniophyllum fasciola.* **M**, habit × ⅔; **N**, flower × 6; **O**, lip × 10; **P**, lip and spur × 8; **Q**, anther cap × 8; **R**, pollinia × 8; **S**, column × 10; **T**, dorsal sepal × 8; **U**, petal × 8; **V**, lateral sepal × 8. A–L drawn from *Cribb & Wheatley* 76 (Kew spirit no. 53090); **M** from *Cribb* 6008; N–V from *Cribb & A. Morrison* 1781 (Kew spirit no. 44186). All drawn by Sue Wickison.

Malleola graeffei (Reichb. f.) P.F. Hunt in Kew Bull. 24: 99 (1970).
Robiquetia constricta (Reichb. f.) Garay in Bot. Mus. Leafl. Harv. Univ. 23(4): 196 (1972).
Robiquetia graeffei (Reichb. f.) Garay, l.c.
Robiquetia mimus (Reichb. f.) Garay, l.c.

Roots terete, white to light green. *Stems* up to 15 cm long, fleshy, covered in leaf sheaths, light green, with brown basal leaf sheaths. *Leaves* alternate, c. 5, oblong, 9–23 cm long, 2–3.5 cm wide, unequally bilobed at apex, light green. *Inflorescence* lateral, 9–20 cm long; rhachis 6–10 cm long. *Flowers* 30–45, non-resupinate, not opening widely, white, pink or dark pink with greenish margins to sepals, petals and apex of lip, with basal flowers developing first; sepals and petals 5–6 mm long; sepals oblong; petals ovate; lip triangular, spur 6–7 mm long, slightly swollen at apex, darker pink; column c. 1.5 mm high. (See plate 33).

DISTRIBUTION: Banks Islands (Vanua Lava), Efate, Epi, Espiritu Santo, Malekula and Pentecost. Also in the Solomon Islands and Fiji.
HABITAT: Sea shore to lower montane forest, abundant, 0–600 m.
COLLECTIONS: *Bourby* 132 (P), 795 & 813 (PVNH); *Bregulla* 6 & 8 (K); *Cabalion* 132, 1840 (PVNH), 1131 & 1529 (P, PVNH); *Cribb & A. Morrison* 1764 & 1799 (K); *Cribb & Wheatley* 122 (K, PVNH); *Green* 1054 (K); *Hallé* in RSNH 6296D, 6296E (P), 6286, 6356, 6393 & 6446 (K, P, PVNH); *Hoock* s.n. (P); *Kajewski* 205 & 448 (K); *McKee* 31971, 32092, 32093, 32672, 43291 (P); *Morrison* in RBG Kew 39 & 42 (K); *Raynal* 15989 & 15995 (P); *Reid* s.n. (K); *Im Thurn* 352 & 562 (K); *Chew Wee-Lek* in RSNH 7 (K); *Veillon* 2409 (P); *Wheatley* 7, 8, 173 & 337 (K, PVNH).

63. **TAENIOPHYLLUM** Blume

Epiphytic. Small leafless plants bearing many long-spreading green roots. *Roots* appressed to the bark of the supporting tree, dorso-ventrally flattened or terete. *Inflorescence* with a short peduncle; rhachis slowly elongating, bearing flowers in succession, 1 or 2 at a time; bracts alternate, in 2 ranks. *Flowers* small; lip spurred; column short; pollinia 4.
In this genus the leaves are reduced to tiny brown scales, the function of the leaves is carried out by the roots, which are green.
A genus of about 100 species from tropical Asia to Australia and the Pacific Islands. A single species in Vanuatu.

T. fasciola (*G. Forst.*) *Reichb. f.* in Seem., Fl. Vit.: 296 (1868).
Epidendrum fasciola G. Forst., Fl. Ins. Austr. Prodr.: 60, n.320 (1786). Type: Tahiti, *G. Forster* 172 (holotype BM, isotype P).
Taeniophyllum seemannii Reichb. f. in Seem., Fl. Vit.: 296 (1868). Type: Fiji, *Seemann* 593 (holotype W, isotype L).

Fig. 33. *Robiquetia bertholdii.* **A**, habit ×⅔; **B**, flower × 3; **C**, column × 10; **D**, pollinia × 10; **E**, stipe × 10; **F**, anther cap × 10; **G**, lip × 6; **H**, dorsal sepal × 6; **J**, petal × 6; **K**, lateral sepal × 6; **L**, longitudinal section of lip and spur × 6; **M**, lip, spur and column × 6. Drawn from *Cribb & A. Morrison* 1799 (Kew spirit no. 43940) by Sue Wickison.

Taeniophyllum asperulum Reichb. f. in Otia Bot. Hamb. 1: 53 (1878). Type: Society Islands, *Wilkes* s.n. (holotype W, syntype W).
Taeniophyllum decipiens Schltr. in Fedde, Rep. Sp. Nov. 9: 112 (1911). Type: Samoa, *Vaupel* 278 (holotype B, isotype K!, W).
Taeniophyllum parhamiae L.O. Williams in Bot. Mus. Leafl. Harv. Univ. 7: 148 (1939). Type: Fiji, *Parham* 3 (holotype AMES).

Roots dorso-ventrally flattened, up to 10–20 cm long, c. 2.5 mm broad. *Inflorescence* short, up to 2 cm long. *Flowers* greenish to pale yellow; sepals c. 3 mm long; dorsal sepal lanceolate; lateral sepals lanceolate, asymmetric; petals falcate; lip slipper-like, c. 1.7 mm long, 1.7 mm wide, semicircular, apiculate when flattened, spur forming a ± right angle with the ovary. (See fig. 31).

DISTRIBUTION: Ambae, Banks Islands (Vanua Lava), Efate, Erromango, Espiritu Santo and Pentecost. Also in the Solomon Islands, the Mariana Islands (Guam), New Caledonia, the Cook Islands, the Horn Islands, Samoa, Fiji, Tonga and Tahiti.
HABITAT: Strand forest, up to 300 m.
COLLECTIONS: *Bourdy* 133 (K, P, PVNH); *Cabalion* 191 (NOU); *McKee* in RSNH 24105 (P); *Nuernbergk* s.n. (K); *Renz* 12537 (G); *Wheatley* 80, 252 & 320 (K, PVNH).

Taeniophyllum trukense Fukuyama, from the Western Pacific area (Caroline Islands) is vegetatively very similar to *T. fasciola*. However, *T. trukense* differs in having symmetrical lateral sepals, lacks a slipper-like lip and has a wedge-shaped spur which is directed backwards.

64. **TUBEROLABIUM** Yamamoto

Epiphytic. Stems short. *Leaves* thick. *Inflorescences* bearing a succession of flowers, or a few at a time. *Flowers* small; sepals and petals free; lip immovably joined to the apex of the column foot, deeply saccate, spurred at the base; pollinia 4.
   A genus of about 11 species from S.E. Asia to the Pacific Islands. A single species in Vanuatu.

T. papuanum (*Schltr.*) *J.J. Wood* **comb. nov.**
Saccolabium papuanum Schltr. in Fedde, Rep. Sp. Nov., Beih. 1: 978 (1913). Types: New Guinea, *Schlechter* 17166 & 18975 (syntypes B, destroyed).
Sarcochilus societatis J.W. Moore in Bull. Bish. Mus., 102: 24 (1933). Type: Society Islands, *Moore* 411 (holotype BISH).
Saccolabium subluteum Rupp in N. Queenl. Nat., 21 (105): 1 (1953). Type: Australia, *Goessling-St. Cloud* 10/1952 (holotype NSW).
Trachoma subluteum (Rupp) Garay in Bot. Mus. Leafl. Harv. Univ., 23: 208 (1972).
Trachoma societatis (J.W. Moore) N. Hallé in l'Orchidophile 40: 1481 (1980).

Roots few, c. 2 mm wide. *Stem* 2–4 cm long. *Leaves* 3–6, elliptic-ovate with an uneven point, 5–15 cm long, 1–1.5 cm wide, light green. *Inflorescence* lateral; peduncle 5 mm long; rhachis c. 1 cm long (elongating as subsequent flowers develop). *Flowers* 1–4; pedicels 3–5 mm long; sepals dull pale yellow with fine brown-red speckles near the base on the interior surface, not opening widely, c. 5 mm long; lip deeply saccate, thick, white with brownish spots at the apex, spur subcylindrical, 1.5 mm long, 1.0 mm wide; column white with a brown-red flash on either side at base.

DISTRIBUTION: Efate. Also in New Guinea, the Solomon Islands, New Caledonia, Fiji, Tahiti, the Tubuai Islands and Australia.
HABITAT: Montane forest, 500 m.
COLLECTIONS: *Cribb & A. Morrison* 1824 (K); *McKee* 31398 (P).

65. **CHRYSOGLOSSUM** Blume

Terrestrial. Pseudobulbs of 1 internode, bearing 1 leaf, or an erect inflorescence, (pseudobulbs often indistinct and plants may appear to consist merely of a rhizome bearing a succession of stalked leaves). *Inflorescence* terminal, racemose. *Flowers* well spaced; sepals and petals free and of equal length; lip trilobed, with a short spur; column with small wings on its front edge, with a conspicuous foot; pollinia 4.
A small genus of about 8 species from S.E. Asia to the Pacific islands. A single species in Vanuatu.

C. vesicatum *Reichb. f.* in Seem., Fl. Vit.: 304 (1868). Type: Fiji, *Seemann* 611 (holotype W).
Chrysoglossum neocaledonicum Schltr. in Engler, Bot. Jahrb. 39: 58 (1906); **synon. nov.** Type: New Caledonia, *Schlechter* 15477 (isotype K!, P!).
Chrysoglossum gibbsiae Rolfe in Gibbs in Journ. Linn. Soc. 39: 175 (1909). Type: Fiji, *Gibbs* 886 (holotype BM).
Collabium vesicatum (Reichb. f.) Schltr. in Fedde, Rep. Sp. Nov. Beih. 1: 98 (1911).
Chrysoglossum aneityumense Ames in Journ. Arn. Arb. 14: 105 (1933); **synon. nov.** Type: Vanuatu, Anatom, *Morrison* s.n. (holotype AMES!).

Rhizome slender, partly concealed by the fibrous remains of sheathing bracts. *Leaf* elliptic, 9–20 cm long, 2.5–4 cm wide, dark green; petiole 1.5–3 cm long. *Inflorescence* up to 22 cm long; peduncle pale green with purple mottling; raceme loosely 5–8-flowered. *Flowers* pale yellowish-green; pedicels slender, c. 1.5 cm long including the ovary, curved upwards; sepals and petals oblong-lanceolate, c. 1 cm long; lip connected to the spur by a narrow claw, trilobed, 1 cm long, 6 mm wide, lateral lobes semi-elliptic, midlobe ovate, disc with 3 calli, the outer calli extending to the centre of the midlobe, the central callus is shorter; column c. 4.5 m long. (See plate 8f).

DISTRIBUTION: Ambae, Anatom, Erromango, Espiritu Santo and Malekula. Also in the Solomon Islands, New Caledonia and Fiji.
HABITAT: Montane forest, 900-1000 m.
COLLECTIONS: *Cabalion* 1678 (PVNH); *Cribb & Wheatley* 45 (K, PVNH); *Hallé* in RSNH 6376 (K, PVNH); *Morrison* s.n. (AMES); *Wheatley* 97 (K, PVNH).

66. **EULOPHIA** R. Brown

Terrestrial. Pseudobulbs few-jointed. *Leaves* plicate. *Inflorescence* lateral, arising from near the base of the pseudobulb, erect, racemose, many-flowered. *Flower* with an entire lip, spurred; pollinia 4.

A pantropical genus of about 200 species. A new genus record for Vanuatu, a single species being recorded.

E. nuda *Lindley*, Gen. Sp. Orch. Pl.: 180 (1833). Type: Malaya, *Wallich* s.n. (holotype K!).
Eulophia squalida Lindley in Bot. Reg. 27: misc. 77 (1841). Type: Philippines, *Cuming* s.n. (holotype K!).
Eulophia macgregorii Ames in Philipp. Journ. Sc., Bot. 9: 12 (1914); **synon. nov.** Type: Mariana Islands, Guam, *R.C. McGregor* 631 (holotype AMES).
For full synonymy see Seidenfaden (1983).

Rhizome up to 15 cm long, sheathed by cataphylls. *Leaves* apical, 2 or more, elliptic-lanceolate, up to 30 cm long, 3.5 cm wide, narrowing at base, with c. 7 veins. *Inflorescence* up to 80 cm long, thick and fleshy. *Flowers* c. 10, greenish-white turning yellow; pedicels c.2.5 cm long; sepals lanceolate, 2.2–2.5 cm long; petals ovate, c. 1.9 cm long; lip ovate to oblong, entire, c. 1.9 cm long, 1 cm wide, spur broad, flattened, c. 3 mm long, pointing downwards.

DISTRIBUTION: Anatom. Widely distributed from Sri Lanka, Asia and the Malay peninsula to Sumatra, the Mariana Islands (Guam), the Solomon Islands, Fiji and Tonga.
HABITAT: Rain forest, 400–1000 m.
COLLECTIONS: *Cheesman* A74 & A73a (BM).

67. **GEODORUM** G. Jackson

Pseudobulbs almost round, subterranean. *Leaves* few, the upper-most largest, broad, petiolate. *Inflorescence* on separate stem, erect; rhachis recurved. *Flowers* not opening widely; sepals and petals similar, the petals broader; lip not spurred, forming with the column-foot a short saccate base with a concave blade; column short with a distinct foot; pollinia 4.

A small genus of 5–10 species from tropical Asia to Australia and the Pacific islands. A single species in Vanuatu.

G. pacificum *Rolfe* in Kew Bull. 1908: 71 (1908). Types: Tonga Islands: Vavau, *Crosby* s.n.; Solomon Islands, *Woodford* s.n. (syntypes K!).
Cymbidium pictum R. Br., Prodr. Fl. Nov. Holl.: 331 (1810). Type: Australia, *R. Brown* 5507 (holotype BM!) non Link (1822).
Geodorum pictum (R. Br.) Lindley, Gen. Sp. Orch. Pl.: 175 (1833).
Geodorum neocaledonicum Kraenzl. in Viertelj. Nat. Ges. Zur. 74: 82 (1929). Type: New Caledonia, *Daeniker* 1384 (holotype Z).

Leaf up to 30 cm long, 7 cm wide, with 5 or 7 principal veins; petiole up to 15 cm long. *Inflorescence* with peduncle up to 45 cm tall, rather fleshy; rhachis 3–5 cm long, bearing a succession of many flowers. *Flowers* pink, mauve or white; sepals, petals and lip c. 1 cm long; sepals acute; lip pale pink with purple stripes and blotches, oblong, bilobed at apex, disc with a yellow-purple U-shaped callus; column c. 5 mm long; column-foot c. 3 mm long.

DISTRIBUTION: Anatom and Erromango. Distributed from New Guinea, New Britain, the Solomon Islands, New Caledonia, Fiji, Samoa, Tonga and Australia.

HABITAT: No information from Vanuatu, (c. 400 m), in the Solomon Islands it is common in secondary scrub and old cultivation areas.

COLLECTIONS: *Cabalion* 943 (PVNH); *Cheesman* A67 (BM); *de la Rüe* s.n. (P).

The name *Geodorum densiflorum* (Lam.) Schltr. has been wrongly applied to this species. A recent study (Jones, 1988) has shown that *G. densiflorum* is a distinct species that is confined to Thailand.

68. **OECEOCLADES** Lindley

Terrestrial. Pseudobulbs few-jointed. *Leaves* plicate or smooth. *Inflorescence* lateral, arising from near base of pseudobulb, erect, racemose, many-flowered. *Flower* with lip trilobed or 4-lobed, spurred; pollinia 4.

A genus of about 15 species in the Old World Tropics, mainly in Africa and Madagascar. A single species in Vanuatu.

O. pulchra (*Thou.*) *Clements & Cribb* in Clements, Cat. Austr. Orch. in press.
Limodorum pulchrum Thou., Orch. Iles Austr. Afr.: t.43–44 (1822). Type: Reunion, *Thouars* s.n. (holotype lost).
Eulophia pulchra (Thou.) Lindley, Gen. Sp. Orch. Pl.: 182 (1833).
Eulophia macrostachya Lindley, Gen. Sp. Orch. Pl.: 183 (1833). Type: Sri Lanka, *Macrae* 27 (holotype K!).
Eulophia novaeebudae Kraenzl. in Bull. Soc. Bot. France 76: 301 (1929); **synon. nov.** Type: Vanuatu, Efate, *Levat* s.n. (holotype MPU).

Pseudobulbs c. 10 cm long, sheathed by cataphylls, and the fibres of decayed cataphylls. *Leaves* plicate, apical, 2 (or 3), elliptic, 18–28 cm long, 4.5–7 cm wide, with 3 strongly marked veins; petiole c. 10 cm long. *Inflorescence* up to 75 cm long. *Flowers* green, with a pale yellow lip with purple lines; pedicel 1.5–2 cm long; sepals ovate, c. 1 cm long; petals narrower, c. 9 mm long; lip trilobed, lateral lobes erect, midlobe much broader than long, broadly cleft, with a divided callus at its base, spur short and spherical.

DISTRIBUTION: Anatom, Erromango and Malekula. Widely distributed from Madagascar, Asia and S.E. Asia to New Guinea, the Solomon Islands, the Mariana Islands, Fiji and Australia.

HABITAT: Rain forest, c. 230 m.

COLLECTIONS: *Hallé* in RSNH 6376 (P); *Morrison* in RBG Kew 147, 148 & 152 (K).

69. **DIPODIUM** R. Brown

Saprophytic. Terrestrial. Roots thick and brittle. *Stems* leafless, sheathed by loosely imbricate bracts. *Inflorescence* erect. *Flowers* widely spreading; lip distinctly trilobed, lateral lobes slender and more or less column-embracing, midlobe much larger than lateral lobes, convex, pubescent in part; column c. half the length of the lip; pollinia 2.

A genus of about 20 species from tropical Asia to Australia and the Pacific Islands. A single species in Vanuatu.

D. punctatum (*J.E. Smith*) *R. Br.*, Prodr. Fl. Novae Holl.: 331 (1810).
Dendrobium punctatum J.E. Smith, Exotic Bot. 1: 21, t. 12 (1804). Type: Australia, *White* s.n. (holotype K!).

var. **squamatum** (*G. Forst.*) *Finet ex Guill.* in Not. Syst. 10(2): 68 (1941).
Ophrys squamata G. Forst., Fl. Ins. Austr. Prodr.: 59 (1786). Type: New Caledonia, *Forster* 173 (holotype K!; isotype P!).
Cymbidium squamatum (G. Forst.) Sw. in Vet. Acad. Handl. Stockh. 21: 238 (1800).
Dipodium squamatum (G. Forst.) R. Br., Prod. Fl. Nov. Holl.: 331 (1810).
Dipodium carinatum Schltr. in Fedde, Rep. Sp. Nov., Beih. 16: 446 (1920); **synon. nov.** Type: Vanuatu, Erromango, *Capt. Braithwaite* s.n. (holotype B).
Dipodium gracile Kraenzl. in Viertelj. Nat. Ges. Zur. 74: 94 (1929). Type: New Caledonia, *Daeniker* 125 (holotype Z).
Dipodium heimianum Kraenzl. in Viertelj. Nat. Ges. Zur. 74: 94 (1929). Type: New Caledonia, *A. Heim* 14 (holotype Z).
Trichochilus neoebudicus Ames in Journ. Arn. Arb. 13: 142 (1932). Type: Vanuatu, Erromango, *Morrison* s.n. (holotype AMES).

Rhizome with an extensive system of long, fleshy roots. *Inflorescence* rather fleshy, 20–100 cm high, varying in colour from green to yellow and brown. *Flowers* 6–50, pink or white, sometimes pale yellow, while tips of sepals may be purple; sepals oblong-ovate, c. 13 mm long; petals slightly shorter and narrower; lip c. 11 mm long, 5 mm wide, lateral lobes oblong, midlobe ovate, disc with 2 pubescent keels. (See fig. 4, plate 8d).

DISTRIBUTION: Anatom and Erromango. Also in New Caledonia.
HABITAT: *Acacia spirorbis* community on volcanic ash, amongst grass and bracken or at the edge of bush, locally abundant, 170–700 m.
COLLECTIONS: *Bernardi* 13342 (K, NOU, P); *Cabalion* 965 (PVNH); *Cheesman* A12a & A14 (BM), 59 (K); *MacGillivray* 932 (K); *Morrison* in RBG KEW 27, 29, 101, 105 & 149 (K); *Raynal* in RSNH 16107 (K, P), 16157 (K, P, PVNH); *de la Rüe* s.n. (P); *Schmid* 3623 (NOU); *Seoule* 73 (P, PVNH); *Chew Wee-Lek* in RSNH 77 (K, P).

This species is currently accepted as a variety of the Australian species, *Dipodium punctatum* (J.E. Smith) R. Br. However, that species is extremely variable and further study may show that the varietal differences cannot be upheld.

ISLAND NAME CHANGES IN VANUATU

The country and the islands have undergone changes in name which are often confusing. The country itself is only properly referred to as the New Hebrides after Cook's voyage of 1774 and as Vanuatu after July 1980. Before Cook, the French explorer Bougainville had called part of it the Great Cyclades.

About 20 of Vanuatu's 80 islands have had more than one name, even now there appears to be uncertainty about some of them. Many of the islands have undergone changes in name, some of them several times, as they were discovered, or rechristened, or reverted to earlier native names in the post-colonial period. Confusion has often been compounded by variations in spelling even when the name itself was more or less agreed upon.

Table 4: Names of Vanuatu islands in present usage and other common variations.

In present usage	Previously
Anatom	Aneityum
Ambae	Aoba
Aniwa	Anina
Efate	Vate, Sandwich
Epi	Api, Tasiko
Erromango	Erromanga, Eromanga
Espiritu Santo (Santo)	Marina
Futuna	Erronan
Gaua	Santa Maria
Maewo	Aurora
Malekula	Malicolo, Malakula, Mallikolo
Pentecost	Raga, Pentecôte

Appendix 2

MAIN ORCHID COLLECTORS IN VANUATU

Collector	Dates	Islands visited	Specimens
George Forster	1774	Tanna Malekula	K, BM
John MacGillivray & William Milne (Voyage of the HMS Herald)	1853	Anatom Erromango	BM, K
John MacGillivray	1858-1859	Anatom Tanna Erromango Futuna	BM, K
Captain George Braithwaite	before 1881	Erromango Efate	B
Rev. Frederick A. Campbell	1872–1873	Anatom Tanna Aniwa Erromango Efate Espiritu Santo	MEL
Rev. F.A. Campbell & Capt. Fraser	1873	Erromango	MEL
David Levat	1883	Efate	MPU
Admiral Fairfax	1889	?	K
Dr Alexander Morrison	June–Aug 1896	Efate Anatom Erromango	K (notes E)
Dr Axford	1901	?	NSW
W.T. Quaife	1903	Espiritu Santo	SYD, K
A. Seale	May 1903	Tanna Anatom Efate	BISH
Rev. MacDonald	1872–1906	Anatom	B
P. Veitch (specimens probably collected by somebody else)	before 1877	?	K, W
Sir Everard Im Thurn	1906	Efate Maewo Malekula	K

Collector	Dates	Islands visited	Specimens
Rev. A. William Gunn	1915	Anatom	NSW
Rev. J.H. Lawrie	received 1916 in NSW	Anatom	NSW
John Layard	1915	Malekula	NSW
S. Frank Kajewski	18 Feb 1928–20 March 1929	Anatom Tanna Erromango Efate Vanua Lava	Arnold Arboretum
Lucy Evelyn Cheesman	1928–1930 1954–1955	Malekula Anatom	BM, K
Oxford University Expedition: Ina Baker assisted by Zita Baker	Oct 1933–Feb 1934	Espiritu Santo	BM
Edgar Aubert de la Rüe	Feb 1934	Epi Ambrym Pentecost Erromango Ambae Espiritu Santo	P
M et Mme E. Aubert de la Rüe	Oct 1935–June 1936	Epi Ambrym Pentecost Erromango	P
Luciano Bernardi	April 1968–June 1968	Tanna Erromango Anatom	G (P, K, NOU)
Royal Society Expedition: Peter Green Nicolas Hallé Chew Wee-Lek Jean Raynal Maurice Schmid Jean-Marie Veillon	1971	Espiritu Santo Malekula Efate Erromango Tanna Anatom	K, P, PVNH, NOU
H. Bregulla	1974–5	Efate Espiritu Santo Tongoa	K
G. Hermon Slade	1975–present	Efate	K

Appendix 2

Collector	Dates	Islands visited	Specimens
Pierre Cabalion	16 Dec 1977– 23 Feb 1986 and 8 June 1988	Espiritu Santo Pentecost Efate Erromango Tanna Malekula Epi Anatom Vanua Lava	NOU, PVNH, P, K, BISH
J. Begaud	1979–present	Efate	K
Genevière Bourdy	21 Sept 1985– 23 April 1987	Ambrym Efate Erromango Espiritu Santo Anatom Malekula Maewo Gaua Vanua Lava	NOU, PVNH, P, K
Jean-Marie Veillon	4 Sept 1971– 15 June 1983	Espiritu Santo Erromango Vanua Lava	NOU, PVNH, P, K
David S. Walsh	28 Sept 1972– 21 Oct 1972	Northern Pentecost	PVNH, NSW (SYD?)
Bernard Suprin	16 Feb 1977 and March– April 1978	Espiritu Santo	NOU, P
Maurice Schmid	23 June 1965– 16 Nov 1974	Aoba Maewo Vanua Lava Pentecost Malekula Efate Tanna Anatom Erromango Espiritu Santo	PVNH, NOU, P
Philippe Morat	12 Dec 1976– 15 June 1983	Pentecost Espiritu Santo Efate Tanna Vanua Lava Aniwa	P, NOU, PVNH

Collector	Dates	Islands visited	Specimens
Phillip Cribb and Alasdair Morrison	1980	Efate	K
Chanel Sam	1981	Efate	NOU, P, PVNH, K
Siri Séoulé	31 Aug 1982 and 18 March 1983	Efate Anatom	PVNH, NOU, P
Phillip Cribb and Jos Wheatley	1988	Espiritu Santo	K, PVNH
Jos Wheatley	Sept 1988– present	Espiritu Santo Efate Ambae Pentecost Vanua Lava	K, PVNH

REFERENCES

Ames, O. (1932). A new genus of the Orchidaceae from the New Hebrides – *Trichochilus*, in Journ. Arn. Arb. 13: 142–144.

Ames, O. (1932). Contribution to the flora of the New Hebrides and Santa Cruz Islands; orchids collected by S.F. Kajewski in 1928 and 1929, in Journ. Arn. Arb. 13: 127–141.

Ames, O. (1933). Additional notes on the orchids of the New Hebrides and Santa Cruz Islands, in Journ. Arn. Arb. 14: 101–112.

Baker, J.R. (1935). Espiritu Santo, New Hebrides, in the Geographical Journ. 85, no. 3, 209–233.

Bellamy, J.A. & Saunders, J.C. (1987). The Vanuatu Forest Resource Survey/Inventory Project. Phase 1 Report. CSIRO.

Cabalion, P., Sam, C. & Seoule, S. (1983). Herbiers de Vanuatu représentés à Port-Vila (PVNH) et à Noumea (NOU)-(comptage), in Naika, Journal of the Vanuatu Natural Science Society, Port Vila, Vanuatu. 12: 12–15.

Cabalion, P. (1987). Vanuatu plant collections in the world. Paper presented at the 14th Pacific Science Congress, held at Seoul 20–30th August 1987.

Cheesman, E. (1957). Things worthwhile. Hutchinson (London).

Chew, W.L. (1975). The phanerogamic flora of the New Hebrides and its relationships, in Phil. Trans. R. Soc. Lond. B. 272: 315–328.

Clements, M.A. (in press). Catalogue of Australian orchids. Australian Orchid Foundation, Melbourne.

Cribb, P.J. (1983). A revision of *Dendrobium* section *Latouria*. Kew Bull. 38(2): 229–306.

Cribb, P.J. (1986). A revision of *Dendrobium* section *Spatulata*. Kew Bull. 41(3): 615–692.

Cribb, P.J. (1987). The genus *Dendrobium* in New Guinea, Australia and the Pacific Islands, in Proceedings of the 12th World Orchid Conference. 210–214.

Cribb, P.J. & Tang, C.Z. (1981). *Spathoglottis* in Australia and the Pacific Islands, in Kew Bull. 36: 721–729.

Cribb, P.J. & Lewis, B.A. (1989). New orchids from Vanuatu in the south west Pacific, in Orchid Review 97: in press.

Davies, S.D. et al. (1986). Plants in Danger: what do we know! pp. 397–398. IUCN, RBG Kew.

de Vogel, E.F. (1969). Monograph of the tribe *Apostasieae* (Orchidaceae), in Blumea 17: 313–350.

de Vogel, E.F. (1988). *Pholidota*. Orchid Monographs 3: 58–65. Leiden.

Dockrill, A.W. (1969). Australian Indigenous Orchids. The Society for growing Australian plants.

Docters van Leeuwen, W.M. (1936). Krakatau, 1833–1933, in Annales Jard. Bot. Buit. 46–47: 1–506.

Docters van Leeuwen, W.M. (1937). The biology of *Epipogium roseum* (D. Don) Lindl., in Blumea, supp. 1, 57–65.

References

Douglas, N. (1986). Vanuatu-a guide. Pacific Publications (Aust.), Sydney.

Dressler, R.L. (1981). The orchids. Natural history and classification. Harvard University Press.

Eriks, M.J. (1988). A partial revision of the genus *Epiblastus*. Unpublished manuscript, Rijksherbarium, Leiden.

Forster, G. (1786). Florulae Insularum Australium Prodromus. Goettingen.

Garay, L.A. & Sweet, H.R. (1974). Orchids of the Southern Ryukyu Islands. Harvard University.

Gillison, A.N. & Neil, P.E. (1987). A feasibility study for a proposed *Kauri* reserve on Erromango Island, Republic of Vanuatu. CSIRO.

Gowers, S. (1976). Some common trees of the New Hebrides and their vernacular names. Dept. of Agric., Port Vila.

Guillaumin, A. (1919). Contribution à la flore des Nouvelle-Hébrides. 1, Primisses de la flore d'Efate (recoltes de M. Levat), in Bull. Soc. Bot. Fr. 66: 267–277.

Guillaumin, A. (1927). Contribution à la flore des Nouvelles-Hébrides. 2, liste des plantes connues, in Bull. Soc. Bot, Fr. 74: 693–712.

Guillaumin, A. (1929). Contribution à la flore des Nouvelles-Hébrides. 3, Supplement aux plantes récueillies par M. Levat, in Bull. Soc. Bot. Fr. 76: 298–303.

Guillaumin, A. (1931). Contribution to the flora of the New Hebrides. Plants collected by S.F. Kajewski in 1928 and 1929, in Journ. Arn. Arb. 12: 221–264.

Guillaumin, A. (1932). Contributions to the flora of the New Hebrides. Plants collected by S.F. Kajewski in 1928 & 1929, in Journ. Arn. Arb. 13: 81–126.

Guillaumin, A. (1935). Contribution à la flore des Nouvelles-Hébrides. Plantes recueillies par M. & Mme. Aubert de la Rüe en 1934 (Phanérogames), in Bull. Soc. Bot. Fr. 82: 346–354.

Guillaumin, A. (1937). Contribution à la flore des Nouvelles-Hébrides. Plantes recueillies par M. & Mme. Aubert de la Rüe dans leur deuxième voyage (1935–1936) (Phanérogames), in Bull. Mus. Hist. Nat. Paris 2, 9: 283–306.

Guillaumin, A. (1938). A florula of the island of Espiritu Santo, one of the New Hebrides; with a prefactory note by the leader of the Oxford University Expedition to the New Hebrides, 1933–34 – John R. Baker.

Guillaumin, A. (1948). La Flore Phanérogamique des Nouvelles-Hébrides, in Ann. Mus. Col. Marseille (1–8): 13–16.

Guillaumin, A. (1956). Contribution à la Flore des Nouvelles-Hébrides. Plantes récoltées par Miss Cheesman, in Bull. Soc. Bot. Fr. 103: 278–282.

Hallé, N. (1977). Flore de la Nouvelle-Calédonie et Dependances, 8: Orchidacées. Museum National d'Histoire Naturelle, Paris.

Holttum, R.E. (1964). Orchids of Malaya (third edition). Government printing office, Singapore.

Hoock, J. (1974). Florule Provisoire des Orchidees des Nouvelles Hébrides. Unpublished manuscript, Noumea.

Inder, S. (1987). Vanuatu, self-reliance in the south seas. Imprimatur Press, Sydney, Australia.

Jones, D.L. (1988). Native Orchids of Australia. Reed Books Pty Ltd., New South Wales, Australia.

Kajewski, S.F. (1930). A plant collector's notes on the New Hebrides and Santa Cruz Islands, in Journ. Arn. Arb. 11: 172–180.

Karig, D.E. (1970). Ridges and basins of the Tonga-Kermadec island arc system, in J. Geophys. Res., 75: 239.

Lavarack, P.S.& Gray, B. (1985). Tropical orchids of Australia. Nelson, Australia.

Lee, K.E. (1974). Royal Society and Percy Sladen Expedition to the New Hebrides, 1971. Collection Data. The Royal Society, London.

Lewis, B. (1989). Orchids of Vanuatu, in Orchid Review in press.

Lewis, B., Cribb, P. & Allkin, R. (1988). Orchids of Vanuatu, a provisional checklist. RBG Kew.

Mallick, D.I.J. (1975). Development of the New Hebrides Archipelago, in Phil. Trans. R. Soc. Lond. B. 272: 277–285.

Mawson, D. (1905). The geology of the New Hebrides, in Proc. Linn. Soc. N.S.W. 30: 400–485.

Merrill, E.D. (1946). Plant life of the Pacific world. Macmillan, New York.

Mueller, F. von (1873). Phytography of the New Hebrides and Loyalty Islands from Mr. F.A. Campbell's collections. Appendix to A year in the New Hebrides, Loyalty Islands and New Caledonia, by F.A. Campbell. George Mercer, Geelong, Melbourne.

Naval Intelligence Division (1944). Pacific Islands. Vol. 3: Western Pacific (Tonga to Solomon Islands), 511–606. Naval Intelligence Division.

O' Reilly, P. (1957). Hébridais répertoire bio-bibliographique des Nouvelles-Hebrides. Publications de la Société des Océanistes, No. 6, Paris.

Rasmussen, F.N. (1977). The genus *Corymborkis* Thou. (Orchidaceae). A taxonomic revision, in Bot. Tidsskr. 71: 161–192.

Reeve, T.M. & Woods, P.B. (1980). A preliminary key to the species of *Dendrobium* section *Oxyglossum*, in Orchadian 6 (9): 195–208.

Reeve, T.M. & Woods, P.B. (1981). *Dendrobium delicatulum* Kraenzl. (section Oxyglossum), in Orchadian 7(1): 18–21.

Rogers, R.S. et al. (1982). The Orchidaceae of German New Guinea, translation of the German text of Die Orchidaceen von Deutsch-Neu-Guinea, by R. Schlechter. Translation R.S. Rogers, H.L. Katz & J.T. Simmons, ed. D.F. Blaxell et al. The Australian Orchid Foundation.

Sanford, W.W. (1974). The ecology of orchids, ed C.L. Withner. In The orchids: scientific studies. Wiley-Interscience.

Schmid, M. (1973). Flore des Nouvelles Hébrides. O.R.S.T.O.M. Centre de Noumea.

Schmid, M. (1975). La Flore et la Végétation de la Partie Méridionale de L' Archipel des Nouvelles Hébrides, in Phil. Trans. R. Soc. Lond. B. 272: 329–342.

Schmid, M. (1978). The Melanesian forest ecosystems (New Caledonia, New Hebrides, Fiji Islands and Solomon Islands). Unesco/UNEP/FAO (1978).

References

Schlechter, R. (1911-4). Die Orchidaceen von Deutsch-Neu-Guinea. Fedde, Repert. Sp. Nov., Beih. 1: 1–1079.

Schumann, K. & Lauterbach, C. (1900). Die Flora der deutschen Schutzgebiete in der Südsee, Leipzig.

Schumann, K. & Lauterbach, C. (1905). Nachträge zur Flora der deutschen Schutzgebiete in der Südsee, Leipzig.

Seemann, B. (1865–73). Flora Vitiensis: a description of the plants of the Viti or Fiji Islands with an account of their history, uses and properties. Reeve and Co., London.

Seidenfaden, G. (1976). Orchid Genera in Thailand 4: *Liparis*, in Dansk Bot. Arkiv 31(1): 61.

Seidenfaden, G. (1978). Orchid Genera in Thailand 6: *Neottioideae*, in Dansk Bot. Arkiv 32(2): 151 & 164.

Seidenfaden, G. (1983). Orchid Genera in Thailand 11: *Cymbidieae*, in Opera Botanica 72: 40.

Seidenfaden, G. (1988). Orchid genera in Thailand 14: fifty-nine vandoid genera, in Opera Botanica 95: 270.

Slade, G.H. (1980). *Dendrobium mohlianum*-A case of pollination by birds, in American Orchid Society 49: 869–870.

Thorne, A. & Cribb, P. (1984). Orchids of the Solomon Islands and Bougainville, A preliminary checklist. RBG Kew.

van Royen, P. (1983). The genus *Corybas* in its eastern areas. J. Cramer, Germany.

Whitmore, T.C. (1973). Plate tectonics and some aspects of Pacific plant geography, in New Phytol. 72: 1185–1190.

Plate 1. **A** Espiritu Santo, Mt. Tabwemasana.   **B** *Macodes sanderiana* (Kraenzl.) Rolfe

C Espiritu Santo, Turtle Bay.   **D** *Macodes sanderiana* (Kraenzl.) Rolfe

E *Goodyera viridiflora* (Blume) Blume   **F** *Pristiglottis montana* (Schltr.) Cretz. & J.J. Smith

Plate 2. **A** *Phaius tankervilleae* (Banks ex l'Her.) Blume   **B** *Calanthe triplicata* (Willemet) Ames

C *Megastylis gigas* (Reichb. f.) Schltr.   **D** *Calanthe langei* F. Muell.

E *Spathoglottis pacifica* Reichb. f.   **F** *S. petri* Reichb. f.

Plate 3.   **A** *Trichotosia vulpina* (Reichb. f.) Kraenzl.

B *Mediocalcar paradoxum* (Kraenzl.) Schltr.

C *Liparis condylobulbon* Reichb. f.

D *L. layardii* F. Muell.

E *Coelogyne macdonaldii* F. Muell. & Kraenzl.

F *Pholidota imbricata* (Roxb.) Lindley

Plate 4. **A** *Appendicula reflexa* Blume   **B** *Ceratostylis subulata* Blume

        C *Glossorhyncha macdonaldii* Schltr.   **D** *Octarrhena angraecoides* (Schltr.) Schltr.

        E *Phreatia paleata* Reichb. f.   **F** *Glomera montana* Reichb.f.

Plate 5.   **A** *Dendrobium seemannii* L.O. Williams   **B** *D. mooreanum* Lindley

         C *D. macropus* (Endl.) Reichb. f. ex Lindley   **D** *D. sladei* J.J. Wood & Cribb

         E *D. polysema* Schltr.   **F** *D. platygastrium* Reichb. f.

Plate 6. **A** *Dendrobium laevifolium* Stapf   **B** *D. delicatulum* Kraenzl.
C *D. mohlianum* Reichb. f.   **D** *D. purpureum* Roxb.
E *D. rarum* Schltr.   **F** *D. calcaratum* A. Rich.

Plate 7.   **A** *Bulbophyllum longiscapum* Rolfe   **B** *Gunnarella robertsii* (Schltr.) Senghas

              C *Eria rostriflora* Reichb. f.   **D** *Pedilochilus hermonii* Cribb & B. Lewis

              E *Bulbophyllum streptosepalum* Schltr.   **F** *Flickingeria comata* (Blume) Hawkes

Plate 8.   **A** *Dendrobium macranthum* A. Rich.   **B** *D. macranthum* A. Rich. × *D. conanthum* Schltr.

C *Microtatorchis schlechteri* Garay   **D** *Dipodium punctatum* (J.E. Smith) R. Br. var. *squamatum* (G. Forst.) Finet ex Guill.

E *Dendrobium conanthum* Schltr.   **F** *Chrysoglossum neocaledonicum* Schltr.

ACKNOWLEDGEMENTS

We would like to thank G. Hermon Slade and The Australian Orchid Foundation for their support; Jaap Vermeulen for help with the genus *Bulbophyllum*; Jany Renz for comments on the genera *Peristylus* and *Habenaria*; Pierre Cabalion for information on the history of plant collectors in Vanuatu; Alasdair Morrison, Geoff Dennis, Mark Clements, Paddy Woods, Peter Green, Martin Love, Peter Boyce, Jeff Wheatley, Jeffrey Wood, Sarah Robbins and Sandra Bell for their help and discussion; Bob Allkin for help and support with the ALICE database; Jos. Wheatley for support in the field and field work; Douglas Malosi, Aaron Hanghangkon, Martin and Miriam Horrocks, Barry and Chris Laing, Keith Shaw, Marianne Cribb, Augustine Tabi, Rueben Maura, Graham Malau, Vira Glenlo, Jacques Begaud, Pat Bochenski, Gabriel Cayrol, Frank King and John Jukes for assistance and encouragement in Vanuatu and New Caledonia; the translation department of the Vanuatu government for the Bislama version of the preface; Hewlett Packard (U.K.) for assistance with computer facilities; the Keeper of the Herbarium, Royal Botanic Gardens, Kew; the Keeper of the Herbarium, British Museum (Natural History), London; the Director of the Conservatoire et Jardin Botaniques, Geneva; the Director of the Laboratoire de Phanerogamie, Paris; the Keeper of the Herbarium, Port Vila, Vanuatu; and the Keeper of the Herbarium, Noumea, for access to material and records.

We are particularly grateful to Sue Wickison for her excellent illustrations and for the enthusiasm she has given to the task. We would also like to thank Joyce Stewart for editing the text and for many valuable suggestions on its format; and Mark Coode for additional editorial help and help with transferring the text to the typesetters.

INDEX TO SCIENTIFIC NAMES

Names in *italics* are synonyms. Species not recorded from Vanuatu but mentioned in the text are preceeded by a short line. Numbers in italics correspond to taxa cited in the text but not described.

ACANTHEPHIPPIUM Blume 50
 papuanum Schltr. *14*, 50
 — vitiense L.O. Williams 50
Aetheria oblongifolia (Blume) Lindley 31
AGLOSSORHYNCHA Schltr. 89
 biflora J.J. Smith 89
AGROSTOPHYLLUM Blume 79
 costatum J.J. Smith *14, 79*, 81
 drakeanum Kraenzl. 89
 graminifolium Schltr. *14*, 80
 leucocephalum Schltr. *14*, 80
 majus Hook. f. 80
 — parviflorum J.J. Smith 81
 torricellense Schltr. *14, 79*, 81
Angraecum purpureum Rumph. 103
Anoectochilus montanus Schltr. 30
Anoectochilus sanderianus Kraenzl. 29
Anoectochilus aff. *roxburghii* Kraenzl. 29
APPENDICULA Blume 82
 bracteosa Reichb. f. *14*, 82
 cordata Hook. f. 83
 dalatensis Guillaumin 83
 polystachya (Schltr.) Schltr. *14*, 82
 reflexa Blume 83
 robusta Ridley 83
 vanikorensis Ames 83
 vieillardii Reichb. f. 83
 viridiflora Teijsm. & Binnend. 83
Asocentrum micranthum (Reinw.
  ex Blume) Holtt. 142

Bletia angustifolia Gaud. 57
Bolborchis crociformis Zoll. & Mor. 46
BULBOPHYLLUM Thouars 123
 atrorubens Schltr. *15, 124*, 134
 atroviolaceum Fleischm. & Rech. 131
 betchei F. Muell. 131
 — cavistigma J.J. Smith 133
 clavigerum (Fitzg.) F. Muell. 134
 christophersenii L.O. Williams 128
 — chrysoglossum Schltr. 125
 — foveatum Schltr. 132
 finetianum Schltr. 131
 levatii Kraenzl. 132
 longiflorum Thouars *12*, 134
 longiscapum Rolfe 125
 membranaceum Teijsm. & Binnend. *15, 124*, 128
 microrhombos Schltr. *15, 124*, 125
 minutipetalum Schltr.*15, 124*, 127
 neocaledonicum Schltr. 132
 nigroscapum Ames 133
 pensile Schltr. 132
 polypodioides Schltr. *15, 124*, 133
 ponapense Schltr. 131
 praealtum Kraenzl. 125
 — radicans Bailey 132
 rhomboglossum Schltr. 127

 samoanum Schltr. 128
 savaiense Schltr. 132
 setipes Schltr. 133
 sp. nov. *15, 124*, 131
 stenophyllum Schltr. *15, 124*, 125
 streptosepalum Schltr. *15*, 128
 — tahitiense Nadeaud 133
 — trachyglossum Schltr. 125

CADETIA Gaudich. 93
 biloba (Lindley) Blume 122
 — hispida (A. Rich.) Schltr. 93
 — lucida Schltr. 93
 quandrangularis Cribb & B. Lewis *8, 11, 14*, 93
Caladenia gigas Reichb. f. 37
CALANTHE R. Br. 51
 angraeciflora Reichb. f. 52
 angustifolia auct. non (Blume) Lindley 51
 chrysantha Schltr. 51
 furcata Batem. ex Lindley 52
 hololeuca Reichb. f. 51
 langei F. Muell. 51
 lyroglossa auct. non Reichb. f. 51
 neocaledonia Rendle 51
 neohibernica Schltr. 51
 quaifei Rolfe 52
 triplicata (Willemet) Ames *12*, 52
  var. *angraeciflora* (Reichb. f.) N. Halle 52
 vaupeliana Kraenzl. 51
 ventrilabium Reichb. f. 51
 veratrifolia (Willd.) R. Br. 52
  var. *cleistogama* Schltr. 52
Callista biloba (Lindley) Kuntze 122
CERATOSTYLIS Blume 78
 cepula Reichb. f. 78
 malaccensis Hook. f. 78
 — micrantha Schltr. 79
 subulata Blume 78
Chamaeanthus laxus Schltr. 137
Chamaeanthus robertsii (Schltr.) Schltr. 137
CHEIROSTYLIS Blume 31
 montana Blume *5*, 31
CHRYSOGLOSSUM Blume 151
 aneityumense Ames 151
 gibbsiae Rolfe 151
 neocaledonicum Schltr. 151
 vesicatum Reichb. f. 151
Cirrhopetalum clavigerum Fitzg. 134
Cirrhopetalum kenejianum Schltr. 134
Cirrhopetalum longiflorum (Thou.) Schltr. 134
Cirrhopetalum thouarsii Lindley 134
  var. *concolor* Rolfe 134
Cirrhopetalum umbellatum (G. Forst.) Hooker & Arn. 134
CLEISOSTOMA Blume 143
 marsupiale Kraenzl. 147
 — montanum (J.J. Smith) Garay 143

Notes

Notes

Notes

Notes

Notes

Notes

Notes

Notes